Data Science in Engineering and Management

Data Science in Engineering and Management

Applications, New Developments, and Future Trends

Edited by

Zdzislaw Polkowski
Sambit Kumar Mishra
Julian Vasilev

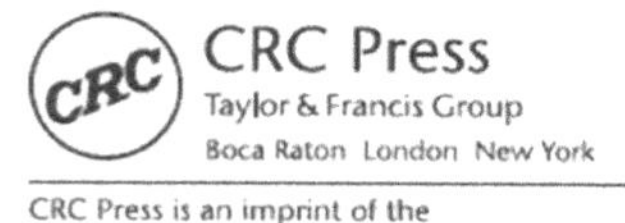

CRC Press is an imprint of the
Taylor & Francis Group, an **informa** business

First edition published 2022
by CRC Press
6000 Broken Sound Parkway NW, Suite 300, Boca Raton, FL 33487-2742

and by CRC Press
2 Park Square, Milton Park, Abingdon, Oxon, OX14 4RN

CRC Press is an imprint of Taylor & Francis Group, LLC

Library of Congress Cataloging-in-Publication Data
[Insert LoC Data here when available]

ISBN: 978-1-032-10625-0 (hbk)
ISBN: 978-1-032-10626-7 (pbk)
ISBN: 978-1-003-21627-8 (ebk)

DOI: 10.1201/9781003216278

Typeset in Times
by Newgen Publishing UK

Contents

Preface

Data science in general is erroneously conflated with big data along with applications to machine learning with large generalized data sets. Therefore, it is highly essential to focus on data sets that are easily analyzable by machines. Techniques as well as data-powered applications are improving day by day. Previously, computation techniques had been responsible for analyzing data sets, but these did not directly lead to data science. This book is aimed at data scientists, analysts, and program managers interested in productivity and improving their business by incorporating data science workflow effectively. It is more focused towards mechanisms of data extraction with classification and architectural concepts and business intelligence with predictive analysis. It covers various problematic aspects of computing as well as data science concepts. It contains information about data science along with its application to business intelligence, prediction, and support for technological innovations. Also it brings insights into data science with applications in general and issues involved in implementation in particular. It is suitable for undergraduate and postgraduate students, researchers, and academicians as well as industry professionals.

Editors

Zdzislaw Polkowski, PhD, is presently associated with the Department of Business Intelligence in Management, Wroclaw University of Economics and Business, Poland. He holds a PhD in computer science and management from Wroclaw University of Technology, a postgraduate degree in microcomputer systems in management from the University of Economics in Wroclaw, and a postgraduate degree in IT in education from the Economics University in Katowice. He obtained his engineering degree in industrial computer systems from the Technical University of Zielona Gora. He has published more than 85 papers in journals and 25 conference proceedings, including more than eight papers in journals indexed in the Web of Science. He has served as a member of the Technical Program Committee in many international conferences in Poland, India, China, Iran, Romania, and Bulgaria. He served as Dean at the Higher Vocational School of Zaglebie Miedziowe in Lubin from October 2009 to July 2012, and was Professor and Rector's Representative for the International Cooperation and Erasmus+ program at Jan Wyzykowski University, Poland from October 2006 to October 2019.

Sambit Kumar Mishra, PhD, has had more than 22 years of teaching experience in different All India Council for Technical Education (AICTE)-approved institutions in India. He obtained his bachelor degree in engineering in computer engineering from Amravati University, Maharashtra, India in 1991, MTech in computer science from the Indian School of Mines, Dhanbad (now Indian Institute of Technology [IIT], Dhanbad), India in 1998, and PhD in computer science and engineering from Siksha 'O' Anusandhan University, Bhubaneswar, Odisha, India in 2015. He has had more than 29 publications in peer-reviewed international journals and is an editorial board member of peer-reviewed indexed journals. Presently, he is associated with the Gandhi Institute for Education and Technology, Baniatangi, Bhubaneswar, Odisha, India.

Julian Vasilev, PhD, is a Professor at the University of Economics, Varna, Bulgaria. Dr. Vasilev has had more than 48 articles published in Scopus-indexed as well as refereed journals. He is also associated with major projects focusing on the development of an innovative methodology of teaching business informatics, IT solutions for business process management, along with the digitization of the economy in an environment of big data.

Contributors

Adithya Ballaji
School of Electrical and Electronics Engineering
REVA University
Bangalore, India

Sasanka Sekhar Bishoyi
Department of Mathematics
Christian College of Engineering and Technology (CCET)
Bhilai, India

Rana Bishwamitra
Department of Computer Science and Applications
MSCB University
Odisha, India

Samarjeet Borah
Sikkim Manipal Institute of Technology
Sikkim Manipal University
Sikkim, India

Ritesh Dash
School of Electrical and Electronics Engineering
REVA University
Bangalore, India

Darshana Desai
Indira College of Engineering and Management
Department of Master of Computer Applications (MCA)
Parandwadi, Pune, India

Marta R. Jabłońska
Department of Computer Science in Economics
Institute of Logistics and Informatics
Faculty of Economics and Sociology
University of Lodz
Lodz, Poland

Hima Bindu Maringanti
Department of Computer Science and Applications
MSCB University
Odisha, India

Antaryami Mishra
Department of Mechanical Engineering
Indira Gandhi Institute of Technology
Sarang, India

Jyoti Prakash Mishra
Gandhi Institute for Education and Technology
Baniatangi, affiliated to Biju Patniak University of Technology
Rourkela, Odisha, India

Sambit Kumar Mishra
Gandhi Institute for Education and Technology
Baniatangi, affiliated to Biju Patniak University of Technology
Rourkela, Odisha, India

Suchismita Mishra
Ambedkar Institute of Technology
Guru Gobind Singh Indraprastha University (GGSIPU)
India

Sudhanshu Bhushan Panda
Department of Mechanical Engineering
Indira Gandhi Institute of Technology
Sarang, India

Miglena Stoyanova
Department of Informatics
University of Economics – Varna
Varna, Bulgaria

Iskren Lyubomilov Tairov
Tsenov Academy of Economics
Svishtov, Bulgaria

Apurva Vashist
Ambedkar Institute of Technology
Guru Gobind Singh Indraprastha University (GGSIPU)
India

1 Concepts for Effective Mobile Device Management in an Enterprise Environment

Iskren Lyubomilov Tairov

1.1 INTRODUCTION

The accelerated development of information technologies (IT) provides serious opportunities for increasing business efficiency. Information mobility is among the leading trends in the field of IT; it significantly improves productivity and efficiency and puts companies into transition. At present, mobile devices have an impressive computing power, they are produced in large quantities, and their popularity is growing significantly faster in all user categories, which leads to their increased importance. They are mainly used for entertainment and personal communication, but in recent years the focus has been set on their key role in achieving business goals by improving the quality of business activities, as well as of people's routine tasks.

By the beginning of 2019, the global forecast was that over 67% or 4.68 billion of the world's population would own and use a mobile phone or smartphone by the end of 2019 (Statista Research Department, 2016).

In a current study as of November 2020, 5.23 billion people own a mobile device; 3.5 billion of these devices are smartphones and 4.78 billion are mobile phones (Statista, 2020) (Figure 1.1). This means that 66.79% of the world's population has a mobile device, making over 10.8 billion mobile connections.

Email: i.tairov@uni-svishtov.bg

DOI: 10.1201/9781003216278-1

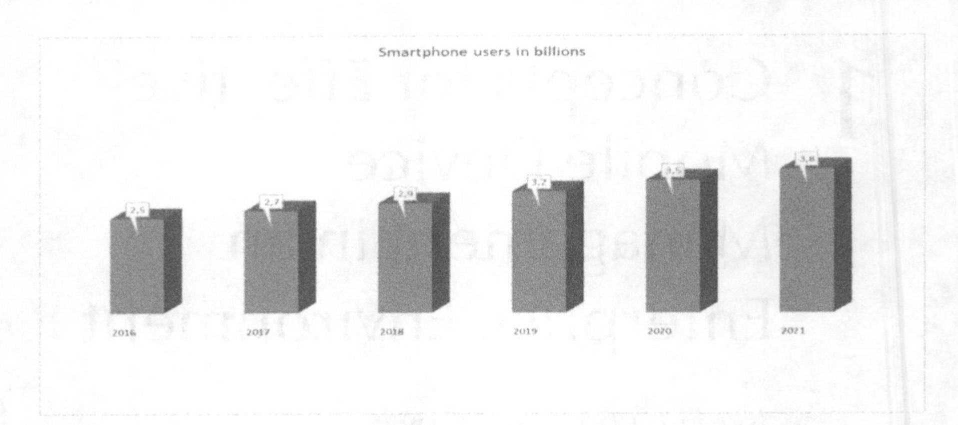

FIGURE 1.1 Smartphone users worldwide. (Adapted from Statista, 2020.)

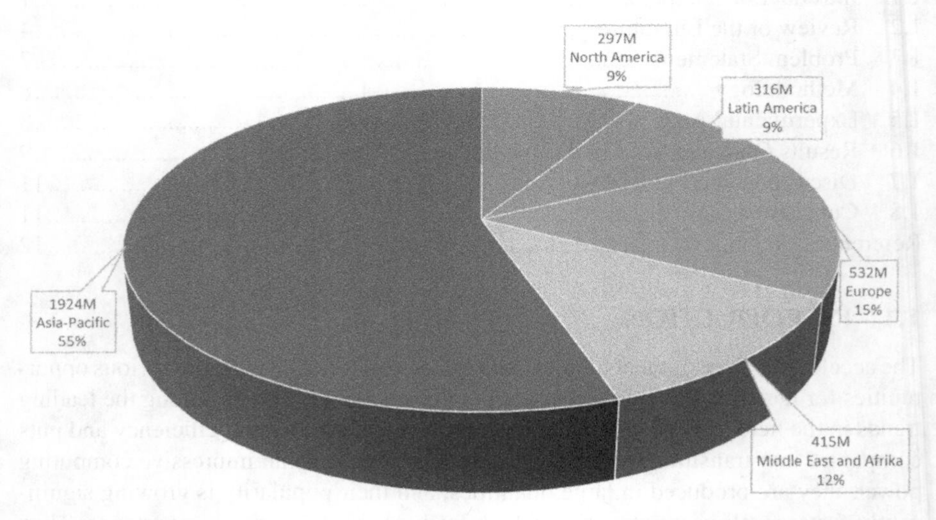

FIGURE 1.2 Smartphone users by region by the year 2020. (Adapted from Chaffey, 2020.)

By region, the most smartphone users are in Asia and the Pacific followed by Europe, the Middle East and Africa and North and Latin America (Figure 1.2).

In detail, China has the most smartphone users, followed by India and the United States (Metev, 2020). The ranking for the top ten countries with the most smartphone users is given in Table 1.1.

China's dominance is explained by the boom in mobile device production in that country. In just a few years, Huawei has reached number 4 after Samsung, Apple and Google Pixel (ICTbuz.com, 2020) and is predicted to overtake Apple in the near future. Chinese manufacturers after Huawei in the rankings include One Plus, Xiaomi and the increasingly popular Oppo.

TABLE 1.1
Top Ten Countries with Regard to Smartphone Use

Country	Users (million)
China	782+
India	386+
USA	235+
Brazil	91+
Russia	84+
Indonesia	67+
Japan	65+
Mexico	60+
Germany	57+
United Kingdom	46+

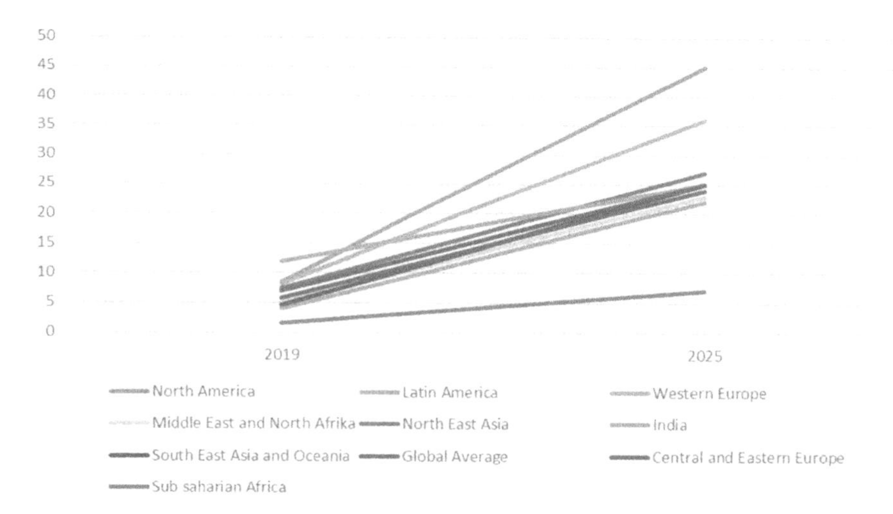

FIGURE 1.3 Mobile data traffic by region measured in GB. (Adapted from Chaffey, 2020.)

Based on the significant number of smartphones in the world, it is clear that a huge amount of data traffic takes place via mobile devices (Figure 1.3). A survey for Ericsson shows data on mobile traffic by region, with a forecast increase of 25% by 2025 (Chaffey, 2020).

The significant number of mobile devices and the presumption that their computing power can be used in the workplace highlight the need to develop strategies for their effective management and control, as well as the deployment of specific schemes aimed at achieving high levels of business process efficiency, as these technologies provide a variety of opportunities to enhance and stimulate business. It is recommended that include elements of the concepts bring your own device (BYOD), choose your own device (CYOD), and so forth, should be included in these strategies. Combinations of several concepts are possible, as well as effective, tools for the management and control of mobile devices, such as mobile device management (MDM), enterprise mobility management (EMM), mobile application management (MAM), and so on, to ensure the successful deployment and management of mobile devices aimed at increasing business efficiency.

1.2 REVIEW OF THE LITERATURE

Different concepts for managing mobile devices in an enterprise environment imply different levels of freedom and control in their use in routine tasks, as well as variability in device ownership. Some of the basic concepts for managing mobile devices in an enterprise environment are described in detail below.

The concept of BYOD involves the use of consumer devices to perform business operations and tasks and subsequent integration of employees' personal mobile devices in the corporate infrastructure (IBM, 2017). These devices include smartphones, tablets, laptops and other wearable devices, finding an application to perform specific tasks, load various applications and access databases. BYOD is driven by the rapid proliferation of mobile devices in recent years and their entry into all walks of life and at the same time is considered one of the leading topics for IT managers today. According to experts from Forrester IT departments are becoming more flexible and more tolerant of the BYOD culture, and at the same time, along with smartphones and tablets, more and more individually owned laptops are used in employees' work (Forester, 2019).

The business benefits of implementing the BYOD initiative generally include increased productivity, collaboration and communication with colleagues and better final results in teamwork and lower costs, especially the cost of mobility, job satisfaction, flexibility and new opportunities for a harmonious combination of personal and work activities by employees. To these are added the creation and maintenance of new channels for interaction with customers, employees, and business partners; more effective cooperation with colleagues and better end results in teamwork; greater efficiency in using IT resources; and improvement of the social climate in the company.

The huge number of devices and the chaos in their use within companies is a major problem in the implementation of the BYOD concept. Other problems are mainly related to the processing of sensitive data, the large range of devices used, and vulnerabilities of the devices themselves (Bluechip, 2012).

Despite these problems, interest in BYOD is constantly growing at a very fast pace and this affects managers and leaders, who cannot remain indifferent to the widespread use of mobile devices. Practice has shown that, in recent years, Chief Information Officers have recognized and supported the importance of workplace mobility. About 77% of IT managers planned to allow staff to use personal mobile devices to access the company's data and applications; 56% supported the strong demand from employees for various mobile devices; 41% of leading staff considered costs to be a critical challenge for BYOD; 30% believed that laptops can be replaced by tablets in the coming years; and almost all IT chiefs expect to provide more than 25 mobility applications over the next few years (Akella et al., 2012).

Because every user has a preference for devices, many prefer their own device in a private environment. In this regard, companies can compile a list of approved devices that will meet the requirements of employees within the company. This is how the CYOD concept was formed. It offers a culturally oriented approach to the mobile business environment; devices can be pre-configured with all the necessary applications for employee productivity and to protect sensitive data that could be present on or accessible through a mobile device.

Focusing on the CYOD concept requires companies to take responsibility for the selection and acquisition of devices for employees (Insight, 2014). Next, a solution must be given to the problems associated with MDM (Networkworld, 2015). MDM can be implemented as software to be pre-installed on devices; however, device ownership issues remain. In this regard, device providers can assist through a number of activities, such as maintaining and optimizing any mobile environment.

In addition to MDM, an effective way to control mobile devices is EMM (Lirex, 2019), which includes a set of technologies, processes and policies aimed at centralized control of the use of mobile devices owned by the organization and its employees. Through the technologies of EMM devices and applications for corporate implementation, use and update can be configured, and also EMM can provide mechanisms for replacement or device access removal to the organization. EMM technologies can be used to track and inventory devices, apply settings in their use, as well as to establish compliance with corporate policies and manage means of access. Basically, EMM solutions are implemented by adding control for data encryption, data access rights, shared devices, packaging applications, and containers and locking devices.

However, practice shows that the application of EMM alone is not enough. It is necessary to implement a set of activities related to MAM (Microsoft, 2020). These activities and technologies are linked to software and services that provide and control access for both employee-owned smartphones and tablets and mobile applications in a business context. These mobile applications can either be commercially available to the public or developed internally within the company. MAM differs from mobile content management (MCM) and MDM in that it focuses on applications that devices use, and not on the management of the devices themselves or the content in their memory. MAM allows a system administrator to have less control over devices, but more control over their applications, while MDM can include both types of management. In general, MAM can be represented as an enterprise application store, which is similar to a typical mobile application store for the purpose of providing updates and adding and removing applications. This also allows the company to monitor how the application works and how it is used. In addition, the system administrator will be able to remotely remove or delete all data from these applications.

All described technologies show the complexity of application of the CYOD concept. Compared to BYOD, it is based on a pre-approved device owned by the organization that can be used anywhere. In most cases, the company purchases, owns and maintains the device and employees have the right to choose from an approved list through user online portals. Challenges here include security and scalability, while BYOD provides the flexibility of any user-owned device to be used anywhere, and users purchase and maintain the device themselves, even installing the applications

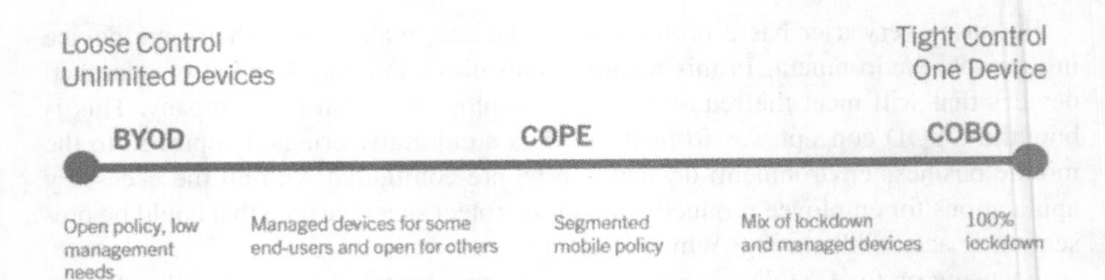

FIGURE 1.4 The corporate-owned, personally enabled (COPE) enterprise mobility concept. BYOD, bring your own device; COBO, corporate-owned, business only.

themselves. Challenges in this concept include data security, experience, consistency, maintenance costs, policy and integration.

The next concept described here is called corporate-owned, business only (COBO), which is largely defined as significantly conservative, because according to this concept the organization owns mobile devices and strictly defines the rules about their usage by employees (BlackBerry, 2015). Applying this concept, companies can introduce a clause banning a smartphone, tablet or other device altogether. It is these drastic restrictions that prove to be the biggest drawback of the COBO approach – the bans introduced encourage employees to carry their personal devices to the workplace and thus to combine work with personal communications from one device. Thus, between the two extremes – the strict policy of COBO and the anarchy of BYOD – a rational alternative appears, called corporate-owned, personally enabled (COPE) (Figure 1.4).

The COPE concept includes the development of a centralized plan, based on which a device is selected from a defined list of pre-configured devices approved by the organization and that are its property, designed to perform business processes and allow personal communications (Black, 2015). From a practical point of view, this concept can impose a high level of mobile management throughout the organization, which is associated with multiple risk profiles of devices, as well as risk relationships, and the result is to achieve solid control through a small number of operating systems, high productivity security and mitigation of information risks without overloading the corporate network.

It is important to note the difference between COPE and CYOD, because they have a similar ideology. What is distinctive about CYOD is the ability for the employee to pay the initial cost of acquiring the device, while the company has the SIM card, the control contract and price reductions and future lower costs.

Another concept for managing mobile devices in the corporate environment is known as company-owned and locked down (COLD) (Brown, 2014). Its essence is the use of devices that are corporate property and are locked, which does not allow employees to change the settings and device applications. Implementing a COLD-based corporate control model can include shared devices, standard configurations, deploying local applications, accessing business-critical host applications, limiting certain device functions and providing protection against device loss. The key to this concept is the strict purpose of the devices for specific corporate purposes, not for entertainment.

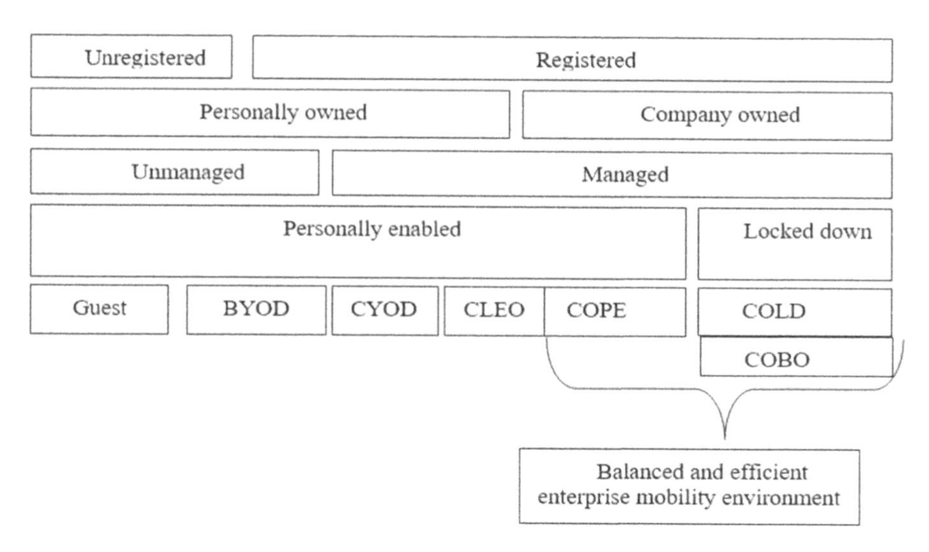

FIGURE 1.5 Forming a balanced and effective approach in the enterprise mobility environment. BYOD, bring your own device; CYOD, choose your own device; CLEO, corporate-liable, employee-owned; COPE, corporate-owned, personally enabled; COLD, company-owned and locked down; COBO, corporate-owned, business only.

The last concept described is called corporate-liable, employee-owned (CLEO). According to this concept, the employee owns the device and the company bears the cost of using it and also the responsibility for management and maintenance. This approach is considered similar to COPE and using it can significantly reduce costs and establish a high level of information security.

In conclusion, of the considered concepts, COBO is the most conservative approach with many prohibitions and restrictions, and in BYOD restrictions are the least. COPE can be seen as an option to soften the conservatism of some COBO policies and tighten control in the BYOD concept, or a liberal approach that combines the strengths of different corporate mobility policies and aims to satisfy employees with minimal information risks and increase the effectiveness of business (Figure 1.5).

1.3 PROBLEM STATEMENT

The evolution of corporate mobility leads to convergence of working conditions, which creates the pre-conditions for building a supportive environment for the introduction of policies based on the effective use and management of mobile devices. In drawing up these policies, three key issues need to be addressed:

1. Concerns about the destructive effect of BYOD in the long run
2. The readiness of a device to be managed and controlled by the company
3. The ability of the selected concept or complex of elements of several concepts to provide an opportunity to impose flexible and detailed mobility policies

that meet the needs of productivity and user usability, without allowing vulnerabilities for the company, information leakage, internal attacks, etc.

Taking a closer look at these issues, it is noted that BYOD is an extremely liberal model for corporate mobility, as it encourages the use of personal devices and applications to perform work-related activities. In this way, employees are more productive, even on weekdays with extended working hours. With the development and duration of a BYOD application in the company, the information becomes more and more vulnerable and there are serious challenges for control and management, security problems, legal risks and more. Thus, BYOD may not be an attractive budget solution and this directs the efforts of different companies in search of effective alternatives.

Next, isolating work and privacy can be a challenge. This can be achieved through the container method, which is to isolate corporate from personal information in the device. In this way, the control of devices through different operating systems can be greatly simplified; in addition, a single model of multiple devices and operating systems can be introduced.

Another problem is the introduction of standard rules for control and management throughout the organization, which reduces the overall costs associated with the complexity of mobile management. Applying different approaches to device variations, combinations of strict rules for actions such as network access, sensitive data and reducing the number of approved devices can be a serious challenge, complemented by attempts to limit the chaos of diverse devices; with this approach, the number of devices can be reduced to a manageable number, thus providing users with a limited but rational choice of device. In this way, IT departments can achieve the desired employee satisfaction while significantly reducing the complexity of controlling devices and applications. Not to be underestimated is the problem of avoiding conflicts and litigation; providing guarantees for the protection of information in personal devices and privacy in case of theft or destruction can reduce the probability of legal action by employees against the organization. Finally, the problem of tightening the control of content by creating procedures that allow close monitoring of corporate content in accordance with regulatory requirements is noted.

1.4 METHODOLOGY

For a methodological basis in the study a systematic approach is used, supplemented by methods such as comparative analysis, synthesis, attitude method, observation method, research approach, inductive and deductive methods and the method of modeling. To achieve the objectives of the study, the statistics of leading research organizations on the use of mobile devices in the workplace in the European Union, the United States and the Middle East were used and compared, as well as the main points on the realities of using mobile devices to perform specific business activities.

1.5 EXPERIMENTAL ANALYSIS

To establish an understanding of the use of mobile devices in corporations, data from several studies showing the use of mobile devices worldwide and by region were presented.

A 2020 Cisco survey showed that 95% of organizations allow personal devices in some way in the workplace (Deyan, 2020). The remaining 5% assume that information security is a high priority and do not allow the use of devices at work.

According to a Samsung survey, 17% of companies surveyed provide mobile devices for their employees, which increases business efficiency by 34%. This is supported by an Apperian study, which found that this indicator was 53%. Regarding implemented device management policies, according to Trustlook, 39% of companies surveyed have a formal BYOD policy, which is slightly worrying in terms of security; the remaining 61% have no regulations in place (Deyan, 2020).

According to Eurostat data, covering EU member states in 2019, more than 50% of employees in the Nordic countries and Ireland are provided with company portable devices that allow mobile internet connection (Eurostat, 2020). The situation is similar in Sweden, with 57% of staff, followed by Ireland (55%), Finland (54%) and Denmark (52%). The situation is radically different in Eastern Europe: only 11% of employees in Bulgaria have been equipped with corporate mobile devices. Numbers are also quite low in Cyprus (16%), Greece (17%), Slovakia, Romania and Portugal, with 18% each.

Another study shows a significant increase in the use of corporate mobile devices for the period 2018–2019. For Sweden, this indicator has increased by 6%, from 51% in 2018 to 57% in 2019. The same situation is observed in Finland, Lithuania and Malta, with the share of employees with corporate mobile devices increasing by 5%.

According to Pew Research, the use of mobile devices in the United States is also high: 69% of employees use their smartphone at work, 80% use it to schedule work and business meetings, and 74% use it to keep notes and diaries of routine tasks and activities (Banda, 2017). A Syntonic study shows that 87% of companies in the United States rely on their employees to use their personal devices to access mobile business applications and services, and almost 50% of companies require their employees to use their personal smartphones (Shields, 2020). About 70% of companies reward and encourage their employees to use mobile devices, but only 29% of employees confirm these practices.

In recent years, the trend of using mobile devices has occurred in Asia and in the Middle East, its presence was quickly felt in the corporate environment. A survey shows that 80% of employees use their own personal devices to work in India (Benq, 2020). The main reason for this is the dynamic change in workplace settings and work culture. With the adoption of the concept of flexible workspaces, the formation of new work environments and models is being considered, including BYOD-oriented approaches.

1.6 RESULTS OBTAINED FROM EXPERIMENTATION

Based on the study, it was noted that companies around the world most often approve of and implement BOYD, adding various ways and tools to achieve a certain degree of control. In greater detail, the results of the study on the use of mobile devices in a corporate environment can be systematized as follows:

- *The number of personal devices used has increased*: Presumably the mobile devices used in the workplace are most often smartphones and tablets. However, the results show a significant change in this situation; a large proportion of the laptops used in the workplace are owned by employees. According to experts from Forrester, this is due to the liberal attitude of IT departments, which has become more flexible and increasingly tolerant of BYOD culture, and at the same time more and more personally owned laptops are used in the work of employees (Gownder, 2019). This wider search for mobile access to companies can lead to a number of important implications, which include investing in expanding remote access to content and data, reviewing the company's application architecture and launching as much software as a service and platform-independent solution as possible, reducing the cost of fixed communications services and channeling the funds released to wireless equipment and services. However, this can exacerbate the problem of control over these devices, which can seriously increase threats to corporate data security.

- *Mobile virtualization on demand is becoming a widely used tool for managing mobile devices*. As many companies that implement BYOD also use technologies to administer mobile devices, this approach is perceived as aggravated and crude. Therefore, alternative methods are being developed to separate personal from corporate data on personal devices, such as mobile virtualized desktop infrastructure, containers, application wrappers and device virtualization tools. The application of such methods in many cases has a negative impact on the user experience and creates risky pre-conditions for the corporate infrastructure and in particular for data. However, the positive effect of mobile virtualization is expected to be felt in the near future based on technologies that can provide an opportunity to enrich policy management across the entire workflow of applications, which will minimize the risks for corporations. It is considered that these technologies will allow the user experience to remain intact and will help to mobilize corporate resources. Ideally, the new technologies in the presented field will be able to fully deal with all the problems associated with BYOD.
- *HTML5-based applications used in corporations are dramatically increasing in popularity*. Many experts, researchers and managers argue that HTML5-based applications are entering the corporate sector at a faster pace than applications created with a special programming language, such as C# or Java. This is driven by the US Federal Communications Commission's intention to expand the wireless spectrum and pursue lower connectivity costs and higher reliability, thereby helping to spread HTML5. Applications based in this technology are simpler and cheaper to develop and maintain, leading to rapid targeting of enterprise applications in the cloud infrastructure, and organizations are expected to increase their costs for cloud applications and infrastructures in the near future. This will make the browser a key tool for device management and is a prerequisite for the development of technologies aimed at secure use of a browser from a mobile device in an enterprise environment.
- *Consumer-oriented mobile services require serious attention to information security*. Devices used in the corporate environment are designed in a way that provides an opportunity to combine information about individuals and their electronic presence, which can lead to serious consequences. It is a fact that employees and companies have recently expressed their concern about the lack

of vaguely established mechanisms governing the use of consumer mobile data. According to many experts, regulators are unlikely to introduce stricter legislation on the interaction of consumer data in the mobile ecosystem. Against this background, it is considered that consumers will become increasingly cautious about the confidentiality of mobile data.

1.7 DISCUSSION AND FUTURE DIRECTIONS

From what has been seen so far, it is considered that, in terms of employees, mobile devices used in the company generally allow for much greater efficiency – people can be productive anywhere. The essential aim is to give weight to the possibility that, from anywhere in the world, individuals can decide on a case and give approval for costs and other activities of this kind. The benefits include no delay when it comes to making a decision with a critical status.

Many companies have deployed the company's customer communication management system on a mobile application that delivers all market information to the company's central system. To implement this approach, companies have established a permanent connection with the software and content of each of the devices, and also provided the opportunity for remote support.

It should be noted that mobile devices and how to control them may vary by sector. In certain sectors there is the opportunity to use mobile devices and applications that allow customers to give feedback related to the status of a particular order or activity. This very quickly confirms a partner, customer or other type of contractor, and the information received is immediately noted in the system, which in turn directly affects other customers of the company.

At present, only some of the key enterprise applications and systems are accessible via mobile devices, and in the future this is expected to be an option for the main enterprise system, which may be an enterprise resource planning class or another system, to run through a mobile application that will greatly increase employee efficiency. A huge number of companies are working in this direction and it is clear that concrete developments will appear in the near future. This in turn will lead to more security challenges.

Another area where companies are focusing their efforts is developing applications to help mobile employees access better IT support services.

1.8 CONCLUSION

Corporate mobility and mobile device management have evolved to such a level that even industries with significant regulations and restrictions have focused on allowing their employees to use company-approved devices for communications and business activities, as well as social media entertainment and networks, games and more. The advantages in the use and management of mobile devices and applications, together with the accelerated growth rate of the number of employees using mobile devices, direct the management to an increased interest in additions or alternatives to the BYOD concept. The expected impact on the organization's ability to implement corporate data security mechanisms located on an increasingly diverse collection of employee-owned mobile devices implies the development of a plan aimed at effective and balanced management of mobile devices. Such an approach implies a focus on the COPE concept. In recent years, the concept has gained increasing popularity due to its ability to combine the freedom of BYOD and the conservatism of the COBO model of corporate mobility. The key point of the COPE concept is that it solves the problem of how to implement effective measures for security of corporate information, without restricting users or efficiency. The concept provides the ability to configure the devices by the company, which is considered an approach to unprecedented control and strict compliance with certain rules in the interest of the development of corporate mobility.

REFERENCES

Akella, J., Brown, B, Gilbert, G. and Wong, L. (2012, 09). Mobility Disruption: A CIO Perspective. Retrieved from www.mckinsey.com/~/media/McKinsey/Business%20Functions/McKinsey%20Digital/Our%20Insights/Mobility%20disruption%20A%20CIO%20perspective/Mobility%20disruption%20A%20CIO%20perspective.pdf

Banda, T. (2017). Mobile Devices 10 Years On. Retrieved 11 13, 2020, from www.netfortris.com/blog/mobile-devices-10-years-on

Benq. (2020, 03 25). *Challenges Presented by Bring Your Own Device (BYOD) in Meeting Rooms*. Retrieved 11 22, 2020, from www.benq.com/en-ap/knowledge-center/knowledge/bring-your-own-device.html

Black, M. (2015, 03 31). *COPE vs. BYOD vs. CYOD – How Should an Enterprise Choose?* Retrieved 10 15, 2020, from www.itbriefcase.net/cope-vs-byod-vs-cyod

BlackBerry. (2015, 06 2). *Beyond BYOD: How Businesses Might Cope with Mobility*. Retrieved 10 13, 2020, from www.bankinfosecurity.com/whitepapers/beyond-byod-how-businesses-might-cope-mobility-w-1684

Bluechip. (2012, 02 02). *The Business Risks and Benefits of Bring Your Own Device (BYOD)*. Retrieved November 12, 2020, from www.bluechip.co.uk/blog/the-business-risks-and-benefits-of-bring-your-own-device-byod/

Brown, S. (2014, 09 29). *What's Your Approach: BYOD, COPE, or COLD?* Retrieved 10 18, 2020, from www.ivanti.com/blog/whats-approach-byod-cope-cold?ldredirect

Chaffey, D. (2020, 09 01). *Mobile Marketing Statistics Compilation*. Retrieved October 28, 2020, from www.smartinsights.com/mobile-marketing/mobile-marketing-analytics/mobile-marketing-statistics/

Deyan, G. (2020, 09 13). *41 Stunning BYOD Stats and Facts to Know in 2020*. Retrieved 11 11, 2020, from https://techjury.net/blog/byod/#gref

Eurostat. (2020, 04 17). *More and More Employees Use Business Mobile Devices*. Retrieved 11 03, 2020, from https://ec.europa.eu/: https://ec.europa.eu/eurostat/web/products-eurostat-news/-/DDN-20200417-1

Forester. (2019, 19 01). *The Future of Enterprise Computing*. Retrieved 11 05, 2020, from www.forrester.com/report/The+Future+Of+Enterprise+Computing/-/E-RES142617

Gownder, J. P. (2019, 01 18). *The Future of Enterprise Computing*. Retrieved 11 14, 2020, from www.forrester.com/report/The+Future+Of+Enterprise+Computing/-/E-RES142617

IBM. (2017). *Bring Your Own Device (BYOD)*. Retrieved 11 12, 2020, from www.ibm.com/security/mobile/maas360/bring-your-own-device

ICTbuz.com. (2020, 11 22). *Top Mobile Phone Brands in the World 2020*. Retrieved 11 30, 2020, from https://ictbuz.com/: https://ictbuz.com/top-mobile-phone-brands/

Insight. (2014). Mobility: BYOD vs. CYOD. Retrieved 11 18, 2020, from https://www.insight.com/content/dam/insight/en_US/pdfs/insight/solutions/cyod-datasheet.pdf

Lirex. (2019, 7 31). *EMM*. Retrieved 11 05, 2020, from https://lirex.bg/home/MDM

Metev, D. (2020, 11 21). *39+ Smartphone Statistics You Should Know in 2020*. Retrieved 11 28, 2020, from https://review42.com/: https://review42.com/smartphone-statistics/

Microsoft. (2020, 9 3). *Mobile Application Management (MAM) Basics*. Retrieved 10 28, 2020, from www.microsoft.com/: https://docs.microsoft.com/en-us/mem/intune/apps/app-management

Networkworld. (2015, 05 12). Mobile device management is becoming more comprehensive. Retrieved 11 14, 2020, from https://networkworld.bg/corp_networks/2015/05/12/3430197_upravlenieto_na_mobilni_ustroistva_stava_vseobhvatno/

Shields, C. (2020, 06 17). *Bring Your Own Device (BYOD) Policy Tips and Best Practices*. Retrieved 10 15, 2020, from www.ntiva.com/blog/bring-your-own-device-byod-policy

Statista. (2020, 11). *November 2020 Mobile User Statistics: Discover the Number of Phones in the World and Smartphone Penetration by Country or Region*. Retrieved 11 15, 2020, from www.bankmycell.com/blog/how-many-phones-are-in-the-world

Statista Research Department (2016, 11 23). *Mobile Phone Users Worldwide 2015–2020*. Retrieved 11 10, 2020, from www.statista.com/statistics/274774/forecast-of-mobile-phone-users-worldwide/

2 Prioritization of Informational Factors and Query Intensification Using a Meta-Heuristic Approach

Samarjeet Borah

2.1 INTRODUCTION

In the current scenario, while considering the virtual platform, the accumulation of data is enhanced with respect to time. Therefore, it is required to monitor data properly along with virtual machines to obtain unique or identical entity types. Also, it will be more ethical to focus on the reality of data predicting patterns as well as the scale of data. Frequently analytical and statistical techniques are more useful in obtaining data variations along with linked information. So data validations along with visualization are more essential during transformation and analytical phases. Again, based

Email: samarjeet.b@smit.smu.edu.in

DOI: 10.1201/9781003216278-2

on hypothetical analysis, in some cases measures can be ascertained on testing of data at multiple stages to signify the desired platforms. The classification techniques must be applied to these testing data sets predicting informational and useful factors. Then it is required to ascertain whether the links are global or local within the data sets.

2.2 REVIEW OF LITERATURE

A number of papers have been published in this domain of research. In this section, some technically impressive works are discussed.

The first work considered focused on a plan enumerator which is associated with the optimizer [1]. In fact, it validates many possible query plans towards implementation of heuristics and minimizes search complexities. As compared with traditional query optimizers, the anomalies linked with some statistical data can be easily retrieved and corrected with the association of suboptimal query plans. Again, Hessel et al. [2] have prioritized deep query learning in which the learning process can minimize overestimation with implementation of decoupling techniques. In fact, this procedure can be similarly reinitiated from the replay buffer.

A discussion on specific stochastic optimization techniques can be found in Krishnan et al. [3] and formulated as a Markov decision process. The main intention in such situations is association of agents towards initiation of a sequence of actions and achievement of optimization criteria. The size of each cell in such situations corresponds to the columns linked to the databases.

In another study, responses associated with join candidates in the optimization process were considered [4]. In this work, investigators initiated actions which might lead to validation of join candidates and the action masking the layer in the neural network was prioritized.

Ryan Marcus et al. [5] focused on developmental growth linked to machine learning in solving the issues associated with the problems. Citing the example, some approaches may be involved in mitigating the issue of cardinality estimation errors towards prediction of the execution time of queries. In their next work [6] they prioritized ReJoin, in which the policy associated with gradient methods can be implemented to obtain better mechanism for join order enumeration; the focus is also on the specific network architecture towards optimization of query plans. The authors also implemented the mechanisms associated with machine learning and were more focused on reinforcement learning [7]. The primary intention in such situations is to optimize specific databases towards enhancement of query execution by using the feedback from past query executions. In specific end-to-end learning optimization applications the implementation of the learned cost model is based on neural networks which can iterate the search algorithm through the large search space of all possible query plans [8].

Immanuel Trummer et al. [9] focused on the Skinner DB system, which uses reinforcement learning for continuous query optimization. Specifically, in such an approach, queries can be dynamically improved based on the bounded quality measure. Another approach focused on the experience replay which is nothing but an accumulator for the observational data associated with a neural network; in fact, it makes the training process more independent [10].

2.3 SCHEMA UPDATE AND MODIFICATION WITH HETEROGENEITY

The applications while analyzing the data sometimes aggregated are and need to be generalized on operators associated with the same. While accumulating of data, non-redundancy features should be checked by matching the primary attributes and should be reliable and consistent with the data. In such cases, data independence plays the major role. It is responsible for modifying the schema of the database concerned at a certain level without affecting the next level. Considering the logical data independence, the logical schema can be updated without affecting the external schema. Similarly, physical data independence has the capability to update the conceptual schema without prior intimation to the internal schema. The integration of schema is really required when provisioned with large-scale heterogeneous data sets. Accordingly, managing queries in such situations can be a difficult task and requires optimization.

2.4 ALLOCATION OF DYNAMICALLY LINKED QUERIES

In many cases, there are provisions to manage the large scale of heterogeneous queries during accumulation from various local databases. In such situations, it is required that the data from various databases can be explicitly associated with the system to help in optimizing queries as well as managing the execution. Now the data analysis portion plays a significant role in retrieving the desired statistical information along with the aggregated data and categorizing them as necessary. The procedure for implementation of time series in this case links to visualization of observational data and resolves issues, if there are any, linked to the data. Again it can be required to acquire the linked patterns of the queries while incorporating the time series data. The coordination between them will have direct access to satisfy the requirements to some extent. There is also provision for generating justifiable interfaced queries to process data from the cloud database to other data sources. In such situations, join indices are initiated to link the databases with each other in the cloud in the virtual platforms. The queries linked with the database in the cloud during the application should have efficient application providers and the services should be hosted in the proper direction in the data centers. In specific cases the large-scale heterogeneous data are confined to acquisition of storage allocation in cyberspace because of multi-dimension and situation complexities. So it is more of a priority to enhance the technologies with these types of data. Also data manipulation operations should be improved to optimize the data or specific queries linked with the data dynamically.

2.5 SIGNIFICANCE OF JOIN ORDERING ALGORITHMS

Based on the market research surveys on innovation, it is clear that the distributed database systems have a number of merits over centralized systems. Of course it is possible to speed up the transition process due to the availability of high-speed communication networks. In fact, real challenges need to be solved before distributed database technology potential can be realized. In such cases, one of the most common issues, i.e. query optimization in large scale distributed databases, sometimes experiences difficulties because of non-deterministic polynomial-hard in nature and can be forced to implement join ordering algorithms using either heuristics or evolutionary techniques. In many cases these experimentations have been proved to be viable methods for obtaining solutions in large-scale distributed database systems. Now considering query optimization, there are strategies to execute queries over the network effectively and to ensure that the complete response time for queries is minimized. Accordingly, it can be required to include the order of relational operations along with access methods for the relations and the algorithms to carry out the operations. In addition, it is necessary to minimize the processor and input/output (I/O) as well as communication costs. Generally, optimizers that are less concerned with complete strategies may be associated with optimal strategy and be cost-effective. The reason is that they are linked with specific operators, i.e. only selection, projection and join enumeration. But it is understood that heuristic database systems resist the combination of relations not linked with any join with the query system. As the operation costs of join operations depend on operand size, it is required to prioritize the same and the mechanisms of every join operation determined. An algorithm for implementation using an iterative approach is stated below.

2.5.1 Algorithm: Implementation Using the Iterative Approach

Step 1: Initialize the vales of alpha, beta and gamma parametric values (alpha = 0.2, beta = 1, gamma = 1).

Step 2: Set the maximum generation = 5000, relation = 20, iteration = 20 and size of query = 5000.

Step 3: Determine the population considering the query size and maximum generation with crossover probability and mutation probability.

Step 4: Determine the intensity of queries prioritizing the relation and size of query.

Step 5: Compare the intensity of queries at different levels and normalize the relation.

Step 6: Determine the optimal query response based on query intensification considering the parametric values.

Step 7: Obtain the maximum response of query intensification based on the relation and size of query.

2.5.2 Intensification of Queries in Databases

Generally, integration and linkages of queries are done to enhance search precision. The main intention in such cases is to regenerate and reconstitute the next-level search technique to obtain more precise results. During initiation of such cases, sometimes difficulties arise with retrieval of data from the cloud as the responses cannot be fully accumulated due to flexibility in query expansion. Therefore, it is imperative to intensify the constraints of the queries to make them more precise and achievable. As reflected in Figure 2.1, when maintaining consistency in queries within the databases and allocating the databases to servers, query intensification time varies depending on the size of data centers (Table 2.1). In many situations, the large-scale sequenced data

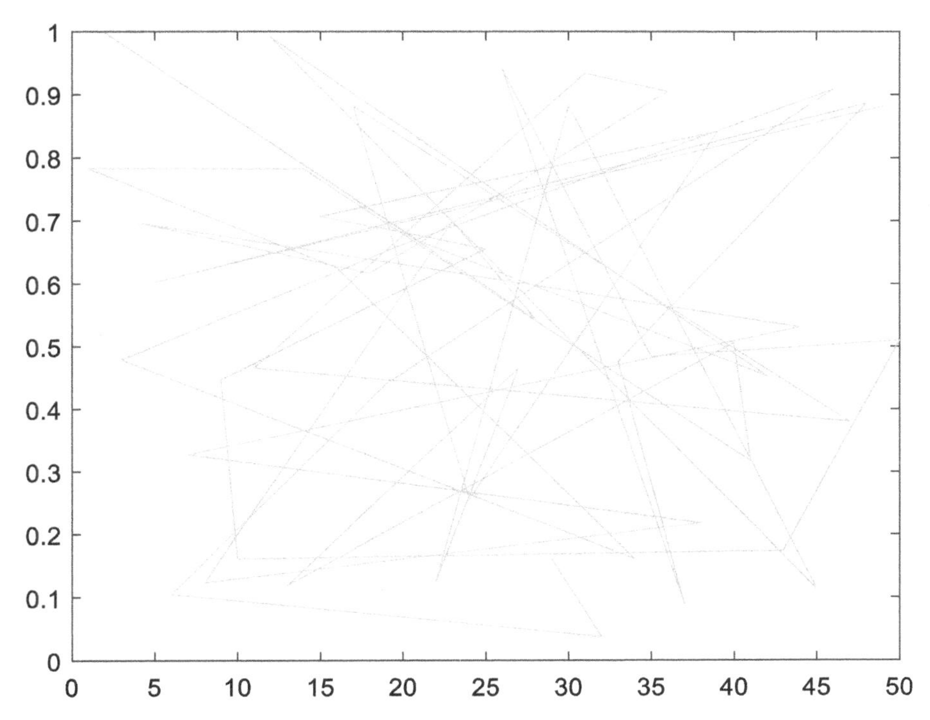

FIGURE 2.1 Data centers vs. intensification of queries in databases.

TABLE 2.1
Intensification of Queries Based on Size of Data Centers

Sl. no.	Size of data centers	Max. allocation of databases per data center	Size of queries within the database	Query intensification (ms)
1	49	20	5000	0.476129890676349
2	48	20	5000	0.474729193479292
3	50	20	5000	0.510114787370240
4	43	20	5000	0.424181223811058
5	37	20	5000	0.4101401908774305

are associated with specified structure and functional variants to operate on selective constraints. To resolve these issues, it is required to implement intensification mechanisms with specified domain structure to enhance the data manipulation capabilities with high-end database triggers. Also this will support the time-scaled data and focus on complete aggregation of data. Further the parameters associated with intensification of queries and large-scale sequenced data can be processed through MATLAB® 13B for further analysis and result generation.

2.6 IMPLEMENTATION OF A META-HEURISTIC APPROACH

The term meta-heuristic is in fact a computational mechanism for addressing complex as well as intractable approaches. Basically it is an iterative process prioritizing the heuristic by encapsulating the linked concepts towards exploration of search spaces to obtain near-optimal solutions. Generally, the strategies associated with this approach guide in the searching process with suitable exploration and to obtain optimal or near-optimal solutions.

2.6.1 Application of Firefly Algorithms

In general fireflies can be linked to each other due to their attractiveness which is proportional to their brightness. Also a less bright firefly can be attracted to a brighter firefly. But the attractiveness of fireflies or their intensity depends on the distance between the fireflies. Of course, with similar brightness fireflies can yield the random movement in the search space. Accordingly, new solutions can be generated randomly. Also attractiveness can decompose linked functional values into smaller scales and obtain a better solution. As such the brightness of the fireflies should be linked with the functional parameters. Practically, the application in such situations is based on global communications among the fireflies. So it is more desirable to obtain global and local optimal communications simultaneously. An algorithm and pseudo-code are discussed in the following section.

2.6.2 Pseudo-Code for Query Intensification Using the Firefly Approach

Step 1: Define the lower and upper parameters and criteria of optimization.
Step 2: Define the fireflies, maximum iterations and maximum generation.
Step 3: Obtain the transportation matrix and size of fireflies.
Step 4: Obtain the parametric value, q = size (transportation matrix, 1).
Step 5: Calculate the fitness parameter focusing the criteria of optimization and light intensity, I.

```
if i < fitness_parameter, then:
  Restore the index parameters
else:
  Sort the fireflies according to light intensity order
```

Step 6: Obtain the updated fireflies and specify the firefly with optimal fitness parameters.
Step 7: Iterate based on parametric values of the fireflies with light intensity values to update the positions.

```
for (t in 1: maxiter) {
  for (i in 1: nrow(fireflies)) {
    for (j in 1: nrow(fireflies)) {
```

```
if(light_intensity [j]< light_intensity [i])) {
transform the firefly i towards j and calculate
the distance between i and j.
```

Step 8: Obtain the new position of the fireflies along with the fitness values.
Step 9: Perform normalization.

```
while (t < maximum generation)
  for i = 1: maxiter
    for j = 1: abs(light_intensity [i], light_
      intensity [j])
        normalize the parametric values
```

Step 10: Obtain the current fitness parameters of fireflies.

2.6.3 Algorithm: Implementation Using the Firefly Approach

Step 1: Determine the cost function.
Step 2: Determine the number of decision variables mentioning lower and upper boundaries, i.e.:

```
number of decision variables = 5, VarMin = -10;
VarMax = 10;
```

Step 3: Define maximum number of iteration, maxit = 1000.
Step 4: Determine number of fireflies, npop = 25.
Step 5: Define the light absorption coefficient, acl = 1, base value of attraction coefficient, acb = 2 and mutation coefficient, mc = 0.02.
Step 6: Determine the uniform mutation range signifying the difference between the decision variables.
Step 7: Compare the decision variables and initialize the firefly structure along with the size of population.
Step 8: Determine the position of initial fireflies along with the cost approximation.
Step 9: Accumulate the feasible and appropriate cost linked to the fireflies.
Step 10: Compare the cost of the fireflies as per position and apply normalization technique.
Step 11: Generate the new solution along with position of fireflies.

2.6.4 Query Response Time and Linkage to Query Intensification

Query response time in a broad sense is linked to measurement of the performance of an individual query. It is primarily considered as elapsed time to initiate and activate the functionality of queries. Sometimes the provision of processing queries is associated with slower execution as the queries are ascertained with variable length and heterogeneity. There are some factors that directly or indirectly affect the query

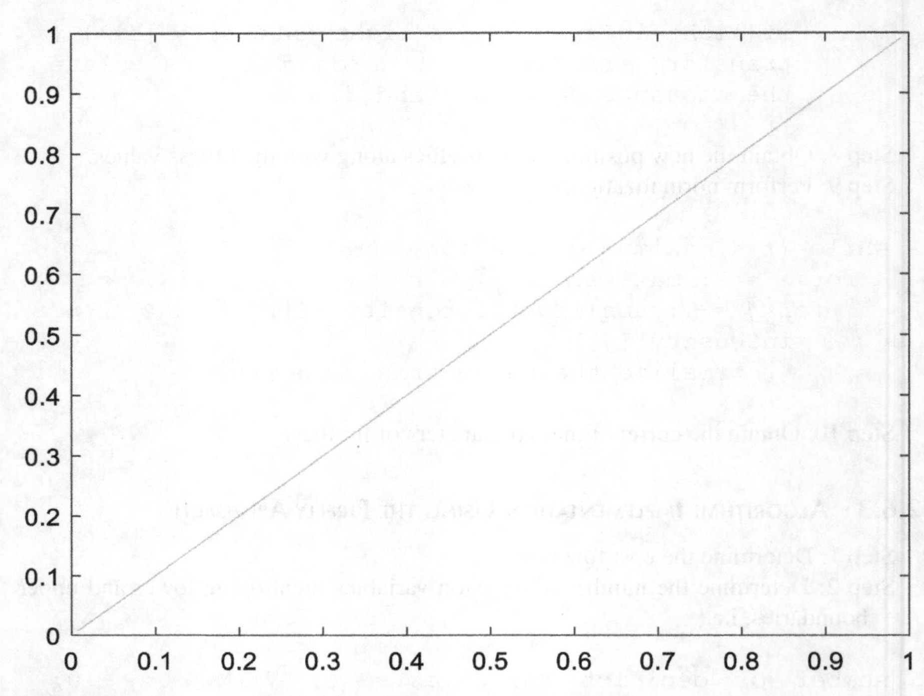

FIGURE 2.2 Query intensification time vs. query response time.

TABLE 2.2
Query Intensification Time with Query Response Time

Sl. no.	Size of queries in database	Query intensification time (ms)	Query response time (ms)
1	5000	0.4	0.31
2	5000	0.6	0.53
3	5000	0.7	0.61
4	5000	0.9	0.89
5	5000	1	1

response time. For example, while accumulating the complete statistics of a database along with allocating the data in memory, the details must be specified to maintain consistency. Similarly, during database partition, the look-up tables of every database must be verified and each term of expansion of the queries examined. When the data are being retrieved from large-scale databases in the cloud, each and every query response should be satisfied based on transformation and query intensification. Based on Figure 2.2, it can be seen that maintaining consistency in the queries, the intensification time of queries is directly proportional to query response time (Table 2.2).

Specifically, the prerequisite data along with the functional parameters linked to query intensification can be processed through MATLAB 13B for further analysis and result generation.

2.7 DISCUSSION AND FUTURE DIRECTIONS

This work, has focused on minimization of search and time complexities and maximum usage of solution spaces. After observation of the mechanism of execution of various heterogeneous queries, it essential to enhance their functionalities. Focusing on the difficulties with traditional retrieval mechanisms, it should be more based on the random selection of parameters and more focused towards obtaining optimality through minimization of solution space as well as time complexities. Also more attention should be given towards satisfaction of each and every query response and maintenance of consistency within the queries.

2.8 CONCLUSION

The mechanism of transformation of data along with the decision parameters have been studied in this work with the intention of obtaining optimal results. Also methodological applications were made on the integral portion of data prioritizing the meta-heuristic approach. The firefly approach was applied in this situation due to its efficiency and outperformance over other conventional mechanisms based on specified computational performances.

REFERENCES

1. Jonas Heitz and Kurt Stockinger. Learning cost model: Query optimization meets deep learning. Zurich University of Applied Science, Project Thesis, 2019.
2. Matteo Hessel, Joseph Modayil, Hado Van Hasselt, Tom Schaul, Georg Ostrovski, Will Dabney, Dan Horgan, Bilal Piot, Mohammad Azar, and David Silver. Rainbow: Combining improvements in deep reinforcement learning. In Thirty-Second AAAI Conference on Artificial Intelligence, 2018.
3. Sanjay Krishnan, Zongheng Yang, Ken Goldberg, Joseph Hellerstein, and Ion Stoica. Learning to optimize join queries with deep reinforcement learning. arXiv preprint arXiv:1808.03196, 2018.
4. Dennis Lee, Haoran Tang, Jeffrey O Zhang, Huazhe Xu, Trevor Darrell, and Pieter Abbeel. Modular architecture for Starcraft II with deep reinforcement learning. In Fourteenth Artificial Intelligence and Interactive Digital Entertainment Conference, 2018.
5. Ryan Marcus, Olga Papaemmanouil. Wisedb: A learning-based workload management advisor for cloud databases. Proceedings of the VLDB Endowment, 9(10):780–791, 2016.
6. Ryan Marcus and Olga Papaemmanouil. Deep reinforcement learning for join order enumeration. In Proceedings of the First International Workshop on Exploiting Artificial Intelligence Techniques for Data Management, page 3. ACM, 2018.
7. Ryan Marcus and Olga Papaemmanouil. Towards a hands-free query optimizer through deep learning. arXiv preprint arXiv:1809.10212, 2018.

8. Ryan Marcus, Parimarjan Negi, Hongzi Mao, Chi Zhang, Mohammad Alizadeh, Tim Kraska, Olga Papaemmanouil, and Nesime Tatbul. Neo: A learned query optimizer. arXiv preprint arXiv:1904.03711, 2019.
9. Immanuel Trummer, Samuel Moseley, Deepak Maram, Saehan Jo, and Joseph Antonakakis. Skinnerdb: Regret-bounded query evaluation via reinforcement learning. Proceedings of the VLDB Endowment, 11(12):2074–2077, 2018.
10. Volodymyr Mnih, Koray Kavukcuoglu, David Silver, Andrei A Rusu, Joel Veness, Marc G Bellemare, Alex Graves, Martin Riedmiller, Andreas K Fidjeland, Georg Ostrovski, et al. Human-level control through deep reinforcement learning. Nature, 518(7540):529, 2015.

3 Estimation and Potential Evaluation of Data Linked to Virtual Machines Using a Computational Approach

Jyoti Prakash Mishra and Sambit Kumar Mishra

3.1 INTRODUCTION

The provision of services linked with virtualized data in different operating systems in general is consolidated to unify as well as integrate machines. In this case, to share the resources physically it is required to link the instances of virtual machines along with their resource capabilities. Sometimes it is necessary to focus on instances to eradicate bugs and provision the instances linked with virtual machines with resource allocation. During these mechanisms, it will be easier to predict the process instantiation and impact on system. The virtualization techniques usually permit operating systems to emulate instructions through translation of codes and sometimes provide multi-platform accessibility. Considering multi-laterality structures, the services linked

Email: jpmishra@gietbbsr.com; sambitmishra@gietbbsr.com

DOI: 10.1201/9781003216278-3

with the cloud should be indulged with optimum expressiveness and multi-tenancy to provision the instantiation. Coincidently, these can be the support mechanisms of the Internet of Things to deploy certain things like identification of data with objects considering the inclusion of static as well as dynamic characteristics, transformation of data in proper specification towards real-time application and computation scenarios of all types of data in datacenters. It is observed that all instances are implemented as real-time applications and each instance in this case is linked with a virtual machine. Therefore the assignment of users to virtual machines is privileged with full accessibility to deploy the application in the network.

Maintaining adequate consistencies in databases is a crucial task. In such cases monitoring the processing of databases implementing all security measures is also important. The strength of a database measured through strong entities should be provisioned with complete accessibility to each and every database activity facilitating the cloud-based database services. In general terms, databases in this case can appear dynamically, being placed in the cloud through which scalability along with redundancies, if any, can be measured properly. The features of the cloud should also be integrated at the infrastructural level prioritizing the abstraction mechanisms on specific issues. As the service provider is provisioned with resource sharing, it is essential to concentrate on virtualization on every database platform. With the significance of the databases along with the associated weight parameters of each database, it is necessary to concentrate on performance of queries linked to databases during the application stage to maximize the utilities of virtual machines periodically consolidated to the cloud.

It is obvious that virtual machines should be provisioned with desired and adequate databases as the implementation of each database server will be on each machine to host many logical databases. Accordingly, the cloud-provisioned mechanism will schedule the accumulation of databases in the machines and incorporate their utilization in the machines. The virtualization of data obtains dynamism while acquiring resources to minimize expenditure and enhance efficiency. It is also known that many educational institutions are facing challenges. Very often it will be necessary to rely on the implementation of virtualization techniques to obtain solutions provisioned with storage as well as computational platforms. As it is linked with distributed computing approaches, it is more essential to focus on the basic requirements. Based on important and critical applications of data in a specific field, it is seen that although classification mechanisms are applied, the confidentiality of data at different stages is uniformly maintained. Of course the database administrator has an important role in such situations. Sometimes in data accumulation as well as the data retrieval process, there may be a threat of security issues. Initiating the process of sanitizing data, i.e. eliminating noise as well as irrelevant facts, will prevent unauthorized access to applications. In addition sensitive data in the cloud should be continually monitoring. While concentrating on new conceptualizations as well as idea generation, it is crucial to prioritize the cloud service provider along with computing platforms and storage allocation. Also the requirement of the infrastructure is essential to meet the needs and demands of organizations, especially educational institutions. Educational institutions that use all these facilities in general have the opportunity to substitute relevant information with the existing datacenters as well as servers and build a distinct architecture in this context. The main intention in this case will be to identify the requirements and simplify the complexities and develop a strong link between the system and service provider to maintain privacy, consistency and integrity.

3.2 REVIEW OF LITERATURE

Naik [1] prioritized the cloud computing infrastructure layer associated with virtual machines provisioned with virtual operating systems. During their research, they observed the concept of virtualization as a confined solution for the support and operations of virtual sessions.

Uhlig et al. [2] focused on the provision of virtual machines and their association with significance as well as the efficiency of I/O systems and computational resources.

Boettiger [3] focused on the basic concepts of signified resources with applications linked to computational nodes. In fact, there should be the provision of a run time environment for applications at operating system levels.

Walters et al. [4] compared full virtualization and para-virtualization concepts and observed that the mechanisms associated with virtualization at operating system level hardly require a hypervisor. Also the operating system will able to be initiated within the same machine.

Koukis et al. [5] focused on a generalized approach towards ensuring the formation of container clusters. They observed that, linked with this approach, each computing cluster can be associated with various nodes, probably virtual servers on hypervisors. In fact, each node can be associated with different containers provisioned with load balancing along with application execution. Sometimes, the platform associated with distributed hardware resources can link to a single pool of resources. In fact, these resources can be used by application frameworks to distribute the load between them.

Apache Software Foundation [6] focused on different platforms like CloudStack, Eucalyptus and OpenNebula associated with solution mechanisms of the classic servers. Observations linked to the solution in virtual environments focus on the performance associated with virtual technologies.

Huang et al. [7] described the solution mechanisms linked to problems of big data. In fact, the requirements in such situations utilize graphics-processing unit (GPU) resources.

Walters et al. [8] focused on the performance of KVM, Xen and VMWare ESXi hypervisors with specified applications. Also the authors prioritized a container management solution for executing specific linked applications. After the experimentation, they observed that KVM achieved 98–100% of base system performance, whereas Xen and VMWare achieved 96–99% of base systems performance.

Bhimani et al. [9] analyzed the effectiveness of virtualization and containerization technologies in obtaining a solution while applying machine learning, large graph processing and query execution techniques linked with specified platforms.

Palit et al. [10] focused on virtualization associated with Xen and KVM hypervisors. Specifically, to provide accessibility, the processing elements should be linked with virtual machines. And as a result, the support technologies associated with the processing elements in the virtual platforms can be provisioned with low overhead costs.

3.3 IMPLEMENTATION USING PARTICLE SWARM OPTIMIZATION

The meta-heuristic property associated with particle swarm optimization (PSO) helps with the initial population of distinct particles, where every particle is linked to the solution. The main intention is to optimize the parametric values and minimize other constraints. In general, particles maintain their position as well as velocity as per the optimal position and pursue optimality linked to the problem domain. The velocity of the particles can be further modified using adopted mechanisms and the parameters can be supported towards obtaining new locational values.

3.4 ACCUMULATION OF MATERIALIZED DATA BY SELECTING AND IMPLEMENTING PARTICLE SWARM OPTIMIZATION

Optimality regarding associated queries can be achieved by minimizing the cost of query processing through selecting and implementing PSO. Accordingly, specific group by-clauses towards linked queries with desired frequencies can be selected to obtain optimal cost parameters of constrained queries during execution. Sometimes complex queries associated with aggregated data are preferable with regard to optimal response time. But as the processing capabilities of large-scale data depend on data aggregation parameters, there may be a chance to make decisions to maintain equilibrium in data storage and cost of aggregated data.

3.4.1 Procedure for Importing and Accessing Data

Step 1: Initiate the connectivity of data imported by activating database triggers.
Step 2: Activate the connection of data specifying additional required options with specified arguments.
Step 3: Analyze the data based upon the stored values and accessibility parameters.
Step 4: Obtain the range of data with maximum tolerance limit.
Step 5: Initiate the operation data = select (Connect, Queryselect, attribute_spec, range) to focus on operation mechanisms on attributes.
Step 6: Reemphasize the connectivity of the database to server by citing the instructions:

```
Sourced_Data = 'Database Server';
Data_connct = database(Sourced_Data, 'x','y').
```

3.4.2 Algorithm to Implement Data Using PSO

Step 1: Initialize the swarm variables and link to query terms of sourced_Data, svr.
Step 2: Determine the range of swarm variables, min_range, max_range.

Step 3: Determine the weight of query terms, wq.
Step 4: Determine the fitness value of swarm variables and link up the swarm variables.
Step 5: Determine the criterion of generation of new position and velocity of swarm variables.
Step 6: Obtain the objective function to compare the values of swarm and determine the best one.
Step 7: Determine the updated swarm velocity.

Initially, connectivity within the databases is established by activating database triggers. After that data are analyzed based upon the stored values and accessibility parameters. Also the swarm variables linked to the query terms are initialized. After evaluating the fitness value of swarm variables, the criteria of generation of new position of the swarm variables are also determined. As shown in Figure 3.1, based on each iterative parameter, the cost of the swarm variables depends on the local parametric values (Table 3.1).

In this work, optimality regarding associated queries is achieved by minimizing the cost of query processing through selection as well as implementation of PSO technique. Based on Figure 3.2, it is clear that the cost parameter of each swarm variable

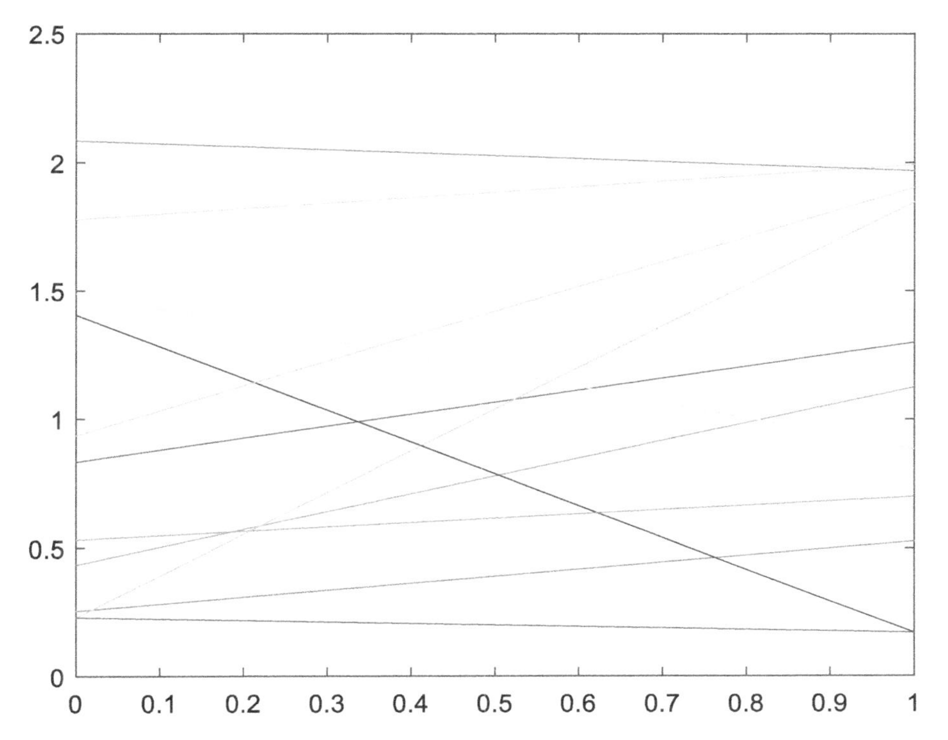

FIGURE 3.1 Cost vs. velocity of swarm based on iterations and local parametric values.

TABLE 3.1
Evaluation of Cost and Velocity of Swarm Based on Iteration and Local Parametric Values

Sl. no.	No. of iterations	Cost	Velocity	Local parameters
1	100	0.434187561569416	0.237420019395760	0.871518878979856
2	100	1.12446918539363	0.709373062848911	0.459716948386581
3	100	0.255771257774236	0.299714165014622	0.762725903490909
4	100	0.528288102107993	0.760957802594520	0.688566403821300
5	100	1.77865674334368	0.960817589452488	0.0776447601243780
6	100	1.99211285093116	0.361885564743106	0.659606406776006
7	100	0.833159216626453	0.995631070089472	0.391865049280808
8	100	1.29745236262446	0.867800203725237	0.236900132758678
9	100	0.230003066795119	0.241595730357075	0.897396820498512
10	100	1.84704635802068	0.527590687767225	0.0953745452536675

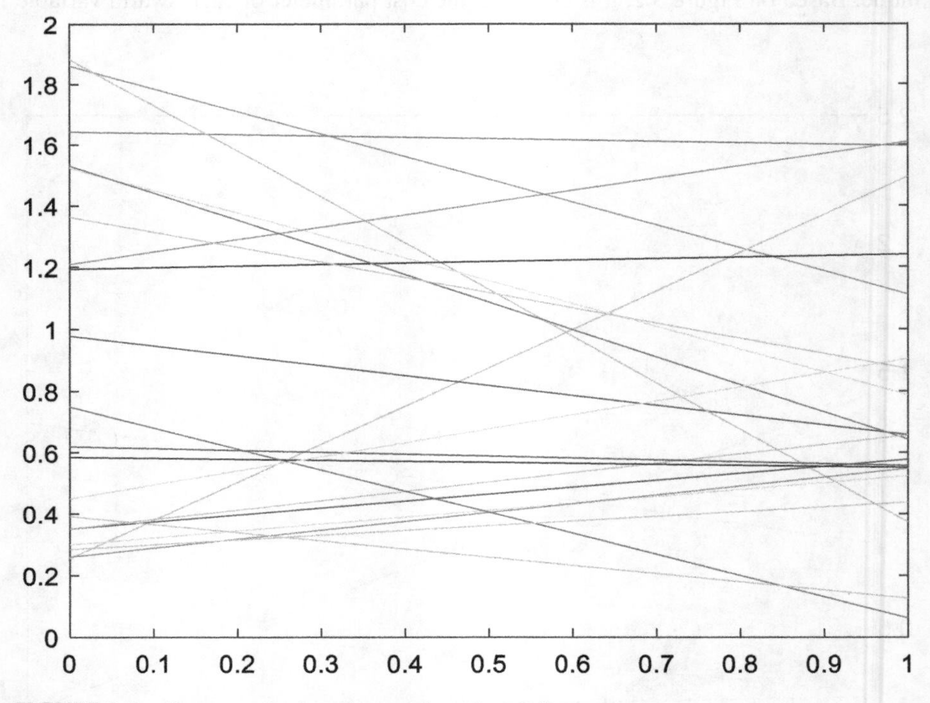

FIGURE 3.2 Cost vs. velocity of swarm based on *R* factor.

TABLE 3.2
Cost and Velocity of Swarm Based on the *R* Factor

Iterations	Cost	Velocity	*R* factor based on swarm	*R* factor based on size of population
100	1.86019710047552	0.351478027547437	0.588637253527382	0.322823395674988
100	1.11202992901444	0.577788096723833	0.181616272270824	0.906229263656453
100	1.64253366451862	0.351245572610408	0.913256477714441	0.266861389803317
100	1.59603260910291	0.656699404380930	0.602160920041822	0.612659638324559
100	0.255192898197977	0.299710133847662	0.0845566767890587	0.707749945073972
100	1.49379181674238	0.523737009571097	0.155389956750810	0.531085935340836
100	0.978585162174092	0.584565190837431	0.490570781019045	0.379617963491418
100	0.657623696252880	0.551470292206574	0.908494076684872	0.429664212746057
100	1.88129850356077	0.259514587691243	0.486807979914986	0.905090546955983
100	0.373509553128401	0.548281387863197	0.555087335542892	0.330216090878052

depends on the *R* factor linked to swarm variable and also the *R* factor linked to the swarm variable is associated with the size of the population (Table 3.2).

3.4.3 Algorithm to Initialize Swarm Variables to Update Particle Position

Step 1: Define the size of the swarm along with the dimension of the problem statement and maximum number of iterations.
Step 2: Determine the cognitive as well as social parametric values and inertia weight.
Step 3: Obtain the initial velocities of the swarm based on the size of population and parametric values.
Step 4: Determine the initial cost of the population implementing the parametric values within the specified function.
Step 5: Obtain the personal learning and global learning coefficient values.
Step 5: Obtain the local minima and global minima of each particle.
Step 6: Obtain the most suitable particle initiating the iteration within the range of maximum range of queries allocated and update the velocity and position of particle.
Step 7: Obtain the new swarm along with associated cost.
Step 8: Compare the cost of the swarm with the specified parametric values. If it is less then update it with the values.
Step 9: Determine the most suitable position of particles applying constriction parameters.

3.5 EXPERIMENTAL ANALYSIS

The experimentation was conducted through MATLAB 13B using PSO technique and comparisons were made on different parameters. It was observed that the effectiveness of PSO is more inclined towards processing the cost of queries. Based on a similar framework linked to query processing associated with dissimilar space allocation, it is clear that with enhancement of the space parameters, the parametric cost of the associated queries almost matches. In fact, the anticipated query-processing result associated with PSO is better compared to other heuristic techniques.

3.6 DISCUSSION AND FUTURE DIRECTIONS

In general, data associated with virtual platform can enhance their efficiency based on demand service implemented through datacenters provisioned with resources. Therefore, it is the prime responsibility of the virtual machine monitor to judge the resources implementing the time-sharing mechanisms. Of course the virtualization in storage has the important role of implementation through storage area networks. Therefore, in this work, prioritization was given to the potentiality of the data linked to the virtual machines using PSO techniques to focus on the parametric constraints as well as the cost of the swarm variables.

3.7 CONCLUSION

The mechanisms associated with virtualization can be implemented to obtain a solution linked to large-scale data. In fact from an analysis of the performance it can be noted that results obtained through deployment of existing mechanisms matter a great deal as compared with hypervisor. But, in specific cases, the associated tasks can be more prioritized on large-scale peripherals. Accordingly, the performance can be compared with the solutions associated with virtual machines.

REFERENCES

1. Naik, N.: Migrating from Virtualization to Dockerization in the Cloud: Simulation and Evaluation of Distributed Systems. In: 2016 IEEE 10th International Symposium on the Maintenance and Evolution of Service-Oriented and Cloud-Based Environments (MESOCA), pp. 1–8. IEEE (2016), DOI: 10.1109/MESOCA.2016.9
2. Uhlig, R., Neiger, G., Rodgers, D., Santoni, A., Martins, F., Anderson, A., Bennett, S., Kagi, A., Leung, F., Smith, L.: Intel Virtualization Technology. Computer 38(5), 48–56 (2005), DOI: 10.1109/MC.2005.163
3. Boettiger, C.: An Introduction to Docker for Reproducible Research. ACM SIGOPS Operating Systems Review 49(1), 71–79 (2015), DOI: 10.1145/2723872.2723882
4. Walters, J.P., Chaudhary, V., Cha, M., Jr., Gallo, S.G.: A Comparison of Virtualization Technologies for HPC. In: 22nd International Conference on Advanced Information Networking and Applications (AINA 2008), pp. 861–868. IEEE (2008), DOI: 10.1109/AINA.2008.45
5. Koukis, V., Venetsanopoulos, C., Koziris, N.: Okeanos: Building a Cloud, Cluster by Cluster. IEEE Internet Computing 17(3), 67–71 (2013), DOI: 10.1109/MIC.2013.43
6. Apache Software Foundation: Apache Mesos. http://mesos.apache.org/, accessed: 2018-12-04.
7. Huang, Q., Xia, J., Yang, C., Liu, K., Li, J., Gui, Z., Hassan, M., Chen, S.: An Experimental Study of Open-Source Cloud Platforms for Dust Storm Forecasting. In: Proceedings of the 20th International Conference on Advances in Geographic

Information Systems – SIGSPATIAL '12, p. 534. ACM Press, New York, New York, USA (2012), DOI: 10.1145/2424321.2424408

8. Walters, J.P., Younge, A.J., Kang, D.I., Yao, K.T., Kang, M., Crago, S.P., Fox, G.C.: GPU Passthrough Performance: A Comparison of KVM, Xen, VMWare ESXi, and LXC for CUDA and OpenCL Applications. In: 2014 IEEE 7th International Conference on Cloud Computing, pp. 636–643. IEEE (2014), DOI: 10.1109/CLOUD.2014.90
9. Bhimani, J., Yang, Z., Leeser, M., Mi, N.: Accelerating Big Data Applications Using Lightweight Virtualization Framework on Enterprise Cloud. In: 2017 IEEE High Performance Extreme Computing Conference (HPEC), pp. 1–7. IEEE (2017), DOI: 10.1109/HPEC.2017.8091086
10. Palit, H.N., Li, X., Lu, S., Larsen, L.C., Setia, J.A.: Evaluating Hardware-Assisted Virtualization for Deploying HPC-as-a-Service. In: Proceedings of the 7th International Workshop on Virtualization Technologies in Distributed Computing – VTDC '13, p. 11. ACM Press, New York, New York, USA (2013), DOI: 10.1145/2465829.2465833.

4 Potential Applications of Blockchain Technology in the Construction Sector

Miglena Stoyanova

4.1 INTRODUCTION

Technological progress and innovations are constantly changing how companies do business to stay ahead in a more competitive market than ever before. Relationships between participants have been significantly transformed by digital platforms. The methods of communication and collaboration have been changed to a large extent. New methods and opportunities for networking are continuously evolving over time.

The rapid advancement of technology has led to the transformation of the construction sector. Traditionally, construction has been inert in the face of changes in its processes and way of functioning, which has led to inefficiency and low productivity (Liu et al., 2017). However, digital development has become a central theme in recent decades for companies to remain competitive in the market (Kamble et al., 2018).

Email: m_stoyanova@ue-varna.bg

DOI: 10.1201/9781003216278-4

Undoubtedly, one of the great modern words that the technological revolution has given rise to is "blockchain". Blockchain is considered to be one of the most innovative technologies of our time. The term has been used more and more frequently in the last few years (Weber et al., 2019). This seems entirely understandable given the fact that the impact of the growing popularity of blockchain technology has been compared to that of the internet in the 1990s (Swan, 2017).

Blockchain technology was originally designed to perform transactions with cryptocurrencies such as Bitcoin or Ethereum (Nakamoto, 2008; Gupta, 2018). The main application area of blockchain is the financial industry (Nofer et al., 2017). But the transparency and security of this technology attract other industries as well. Blockchain platforms are convenient in that they directly connect the client and the contractor without the participation of intermediaries. The technology decentralization allows all operations to be carried out with maximum responsibility of the parties and a guarantee of compliance. These opportunities may be useful in the construction sector, where there are constant problems with timing, payments, etc. Dealing with them obviously requires increased productivity, efficiency and overall process optimization.

Technological innovations in the construction sector may enable reduced administration, real-time planning, improved tracking of goods, efficiency throughout the supply chain and financial flows. New systems, automated processes and reduced manual labor may increase supply chain management efficiency through improved predictability, planning of goods and accurate inventory. These improvements may be achieved using a blockchain solution. Research by Penzes (2018) supports this statement, showing that blockchain can improve ordering and sales processes and facilitate complexity and fragmentation. Another study (Tapscott & Tapscott, 2017) found that blockchain technology can reduce the need for intermediaries, automate processes and improve payment systems.

The purpose of this research was to survey blockchain technology and its potential applications in the construction sector. It examines the basics of blockchain, its current advances and categorization. In addition, the study explores the potential use of blockchain to improve the automation of construction processes and discusses future research areas.

4.2 REVIEW OF LITERATURE

Many researchers discuss the basic blockchain technology concepts, functionality, security, current implementation and transaction protocols (Crosby et al., 2015; Swan, 2015; Zheng et al., 2017; Feig, 2018; Puthal et al., 2018; Brakeville et al., 2019). Blockchain fundamentals, potential applications and new frameworks for distributed ledger technology are reviewed by Wang et al. (2017), Coyne et al. (2018), Li et al. (2018) and Cong et al. (2019). The benefits of using different types of blockchain are studied by Mason et al. (2017), Mathews et al. (2017) and Hammi et al. (2018).

A blockchain is a decentralized database which chronologically and securely records transactions. A transaction can be of cryptocurrency like Bitcoin. However, blockchain transactions can further represent the transfer of value on systems like

Ethereum and others. The value can be a service, a product or an approval in the form of a smart contract.

The blockchain technology is widely defined as a peer-to-peer network (that exists between all organizations involved in a project) in which assets are stored and shared in a distributed ledger, thus eliminating the need for intermediates (Nakamoto, 2008). The assets can be tangible (for example, materials) or intangible (for example, intellectual property, copyright) and consequently affect the blockchain architecture (Gupta, 2018). Unlike a traditional and centralized single-owner system, the infrastructure is owned by all network participants. Each connected computer is called a node. All nodes are equally responsible for the shared database and are peer-to-peer replications. A peer-to-peer replication can be every participant acting as a publisher and subscriber. Using this method the ledger is updated whenever a transaction is made. Each node in the network contains a complete copy of the entire ledger – from the first block (called the genesis block) to the latestone.

The blockchain technology is a digital data structure and a complete ledger of all historical transactions, creating the "single source of truth". The transactions are aggregated into *blocks* that are ordered chronologically. In addition to storing the actual *transaction data*, each block also contains a *hash value*, generated using a cryptographic hash algorithm (most commonly used is Secure Hash Algorithm-256), a *hash value of the previous block* (the parent block), a *timestamp* displaying the time the block was created, a *public key* signed with its corresponding private key to secure and validate the user and a cryptographic *nonce*. Nonce is a 4-byte field that usually starts from zero and is incremented by one for each hash calculation, acting like a transaction counter. The sequence of hashes, connecting each block to its predecessor, generates a chain (hence the name blockchain) that leads to the first block – the genesis block.

Three generations of blockchain technology are distinguished in the scientific literature:

1. Blockchain 1.0 covers applications that allow transactions with digital cryptocurrencies.
2. Blockchain 2.0 expands blockchain 1.0 by introducing smart contracts and a range of applications beyond cryptocurrency transactions.
3. Blockchain 3.0 improves the capabilities of blockchain 2.0 in terms of transaction time, scalability and ease of execution using decentralized applications.

A blockchain is characterized by the following features: decentralization, immutability, anonymity and auditability (Zheng et al., 2017; Andoni et al., 2019).

Decentralization relates to the possibility of the network transactions to be performed between each node without the need for a central trusted participant (for example, banks). Therefore, everyone can participate in the consensus (trust) process (the process of establishing agreements among the mistrusted participants of a blockchain network through cryptographic codes) and the power is decentralized among all participants.

Immutability refers to the process by which all transactions have to be confirmed, recorded and distributed in blocks on the network. In addition, all blocks have to be validated by other nodes to be added to the blockchain. This makes it almost impossible to forge.

Anonymity means that users may avoid revealing their identity. Each user joins a network with a generated address that can be generated repeatedly and therefore none of the other participants can know the actual identity of the sender. Thus, participants are considered as pseudo-anonymous.

Auditability or traceability is another useful feature of blockchain technology. Since the transactions in each block are time-stamped, validated and chained to each other, this allows users to trace a transaction to a previous transaction. Therefore, it improves the traceability and transparency of data stored in the blockchain.

After the first introduction of Nakatomo's technology (blockchain 1.0), various other configurations of blockchains (blockchain 2.0 and 3.0) with different characteristics were developed. In addition to the original public permission-less blockchain, other blockchains have appeared that are permissioned. This means that there are restrictions for participation in the network and access to the shared ledger is limited to specific users. Therefore, these permissioned blockchains have participants who are responsible for the blockchain. Two configurations exist in this case – a private blockchain or a hybrid version of the previous two, called consortium blockchain. Since the blockchains are programmable, there are lots of versions in addition to the mentioned configurations (Maxwell & Salmon, 2017). The three most common are presented in Figure 4.1.

Blockchain technology has lots of areas where it can add value today. Depending on the context, requirements and purpose of use of the blockchain, each type of the three mentioned blockchain configurations has advantages and disadvantages. Although the different approaches for implementing blockchain technology may have the same common features, different blockchain platforms have different technology and security risks. Organizations have to evaluate the various solutions throughout their lifecycle to be sure they meet their specific needs.

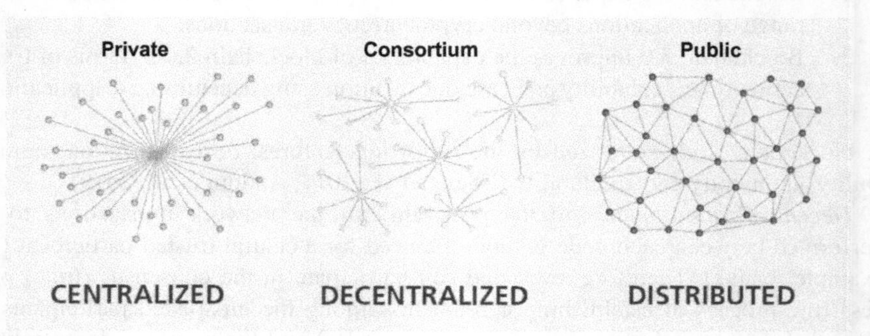

FIGURE 4.1 Various blockchain configurations. (Adapted from Business Blockchain HQ. Blockchain Fundamentals.)

As discussed earlier, the basic principle of blockchain technology is the creation of a decentralized peer-to-peer network with a shared and distributed database where users can be anonymous. In addition, all participants reach consensus in order to verify the transactions and prevent tampering. Thus, the need for intermediates and a central authority is eliminated. The blockchain enables companies to do business with each other more easily by using shared business processes encoded in a common platform. This ecosystem-based approach allows organizations to buy, track and pay for goods and services much faster and more efficiently.

4.3 OVERVIEW OF BLOCKCHAIN IN CONSTRUCTION

A lot of professionals are involved in the construction process. They need to exchange information for successful design, implementation and project operations. Many intermediaries are used for authentication of the whole process – bankers, regulators, lawyers, insurers, etc. It is necessary to build trust between all stakeholders. The transition from traditional methods to digital modes of operations makes it possible to adapt a digital tool like blockchain that can encourage and give confidence among participants.

Construction companies which want open, reliable Internet of Things (IoT) communications without the need to rely on intermediaries can use a private blockchain as a solution that enables data security between IoT devices. The strength of the blockchain is that it sets very powerful standards which are easy to be adopted and do not interfere with the existing processes.

Blockchain manages trust systematically to ensure a level of transparency that increases productivity and simplicity. Blockchain allows contractors dealing with many different subcontractors, who use different systems, to optimize and manage all the data in one immutable ledger. The construction sector has some of the highest costs for litigation of any industry. Thus, the idea of partners coming together and solving problems through everyone entering the same database is an appreciated development. Blockchain implementation guarantees that all data, regardless of source or system, is verified and reliable. Its leading goal is to break down data silos in order to use real-time information. Blockchain technology is an incentive for the construction sector as it is a shared and open platform for securing members who foster trust, transparency and integrity.

Construction project data is usually scattered, comes from different sources and is difficult to verify. This is complemented by the lack of project cooperation and the silos, which are developed using different technological solutions. The main problem of the siloed approach is that it often produces inaccurate or wrong data. Data that is stored in isolated systems is usually fragmented and hardly shared between organizations. Thus, it loses its value and possibility for verification.

The combination of blockchain and IoT aims to improve performance in the construction sector. Blockchain creates a reliable chain of transactions, events, assets and critical details for the projects. This allows information from different systems (e.g., project management systems, accounting systems) and emails to be collected in one place. Hence, a verified record of all transactions is created in the blockchain, which ensures that data cannot be lost. This means that there is one shared version of the truth about each project. This is necessary for the construction sector to provide data integrity, eliminate duplication of information and reduce errors. Blockchain cannot replace the various software applications. It just allows them to work closer and facilitates operational processes.

Currently, the implementation of blockchain has not gained support in the construction industry. The sector and participants lack confidence in its ability for change. Moreover, knowledge about this technology in the sector is low (Mason & Escott, 2018). Although the existing research in the field of blockchain in the construction sector is scarce, it shows that there is potential. The application of blockchain allows easy collaboration between participants, separation of projects, automatic settlement and clearing of transactions (Hughes, 2017).

One study that deals with blockchain technology in the construction sector was carried out by Lanko, Vain and Kaklauskas (2018). The authors explored the use of radiofrequency identification (RFID) technology along with blockchain technology, including the potential of smart contracts to improve the transport of ready-mix concrete. These improvements refer to increased efficiency in the supply chain, improved trust between participants and the potential to increase the automation of ready-mix concrete transportation. The authors considered that a globalized decentralized system (global blockchain) will significantly increase efficiency and automation, reduce misinformation and loss of human influence and eliminate trust issues in the supply chain. This effect has been observed upstream and downstream in the construction supply chain.

Another study (Mason & Escott, 2018) examined smart contracts within the construction sector in the UK market. The researchers found that construction companies have very low levels of knowledge and confidence about the usefulness of the technology. Mason and Escott concluded that there is a need for maturity in the technology for it to become useful. In addition, Mason's study (2017) identifies nine parameters that need to be overcome to enable the application of smart contracts in construction. The most important are clear definition, identification of unsatisfactory areas in the construction sector and development of a common strategy for technology implementation.

The adoption of blockchain technology in the construction domain is expected to increase both business competitiveness and overall system efficiency while achieving synergy and improving communication protocols between the entire network and streamlining payment systems (Ablyazov & Petrov, 2019). An example of blockchain application in construction is outlined in the Agreements of Participatory Construction Interest (APCI) project. The APCI network is used to register apartment purchase agreements using a consensus, agreed by applicants and stakeholders.

4.4 POTENTIAL APPLICATIONS OF BLOCKCHAIN TECHNOLOGY IN CONSTRUCTION

Although still in its infancy, blockchain technology has the potential to accelerate and optimize much of today's design and engineering practices with many benefits for the company, sector, customers and society. The adoption of such technology can also help in a more rational exchange of information, which is crucial for the success of construction projects. This technology can revolutionize how project teams build in the future. Blockchain implementation can lead to effective management and use of several tools that drive efficiency, transform the culture of the sector and advance its future progress (Mathews et al., 2017), such as:

- Building Information Modeling (BIM) software for intelligent and collaborative 3D design and modeling
- Cloud technology, allowing the creation and coordination of a visualized database in real time and serving as a platform for multidisciplinary cooperation
- Smart contracts – A set of coded instructions that can be automatically executed when certain conditions are met
- Reality capture technology, allowing verification and conversion of digital assets into real value
- IoT management – The distributed wireless sensor networks play a vital role in construction projects and the blockchain architecture can improve IoT by minimizing scarcity and exploiting its potential

Blockchain technology can significantly reduce administrative costs, effectively protect intellectual property rights and eliminate paper documents, manual verifications and contract execution. Design professionals can achieve potentially new revenue through the evaluation and sale of designs and workflows.

Reputation building is an important asset for any organization that is difficult to determine and compare. Blockchain can facilitate the creation of a register, consisting of previous achievements and qualifications. It can enable the comparison of team constellations and may assist in the decision-making processes for clients and project managers to select a well-balanced team with a diverse sets of skills, experience and versatility (Mathews et al., 2017).

4.4.1 Smart Contracts

Smart contracts are largely considered as the future of the construction sector. According to Mason and Escott (2018), automated contracts, which reduce the need for intermediaries and save time and money, are ready to leave their mark on the sector. Blockchain is one way to update these contracts and record transactions.

Smart contracts are essentially self-executed computer programs that work on the *if–then* principle and self-enforce independently the terms and conditions set out in legal agreements. This is how contracts are administered. For example, *if* a contractor has reached an agreed stage in building construction, *then* the contractor wants it to be inspected. *If* the authority responsible for inspection of the work approves it, *then* the contractor receives payment. Smart contracts can be used for such *if–then* scenarios and securely recorded in the blockchain.

Another example is the delivery of materials or goods. Smart contracts allow customers to buy directly from the supplier, because they can provide more trust in the transaction. Payment to a supplier can be made in stages and the responsibility can be transferred to different parties. Let's take as an example some kind of construction equipment. The customer can purchase the equipment directly from the supplier, pay part of the costs after it is verified that the equipment has left the place of origin, transfer responsibility to the shipping company, release additional payment when the equipment arrives on site and transfer responsibility again, this time to the contractor responsible for the installation. The final payment can be made after the installation and commissioning of the construction equipment. All these stages can be stored in the blockchain and provide more opportunities for direct transactions without the need for intermediaries.

A large variety of companies are usually involved in a construction project. The company that manages the project usually has to wait to receive contracts from all these companies, as well as change orders and additions. It has to pay all participating companies and wait for their status reports. By contrast, blockchain and smart contracts first create a centralized tracking system. The participating companies set the rules, regulations, deadlines and sanctions around the project on which they are working together (this is the part of the smart contract). The blockchain then works to automatically enforce these obligations as the project progresses. For example, if a material is not delivered on time, the blockchain system records that and an appropriate action may be imposed, according to the pre-agreed rules and regulations. If this is a penalty fee, for example, the transaction can be done automatically via Bitcoin or cryptocurrency.

Blockchain also has the potential to thoroughly improve the global construction supply chain. Blockchain automatically captures where the project's assets are at any time. It identifies the state of an asset and who has custody. For example, steel that moves through a construction supply chain can be tracked and controlled in detail automatically. All relevant information is available to all stakeholders in the peer-to-peer network. Blockchains free projects from being driven by documents to being driven by data.

The benefits of this type of visibility accumulate up and down the construction supply chain. The implementation of such a system can not only improve the efficiency and resolution of disputes, but can also work to increase the accountability of the parties involved. Inventory management is simplified and ordering and receiving goods are only possible according to current needs (also known as "just-in-time planning"). This kind of management significantly reduces waste. All this contributes to lower costs and faster schedules, which are the two biggest drivers of value in the construction sector.

The potentials and benefits of blockchain technology and smart contracts application in construction sector can be summarized as follows:

- If the terms and conditions of a smart contract are precisely registered, their execution and monitoring are very accurate.
- Every business interaction, transaction, payment and execution can be registered in a blockchain, which makes the whole process traceable and transparent.
- The network of smart contracts can provide transparency and reduced complexity for the entire construction procurement. Therefore, the risk of late payments and the number of conflicts can be reduced.
- Major cost savings can be achieved with overhead costs, administration and control of projects. In addition, project procurement information is recorded

in a traceable way, which gives insights for project estimation and cost optimization.

- Contractual cooperation, which is maintained and automated by smart contracts, can significantly reduce the number of claims and disputes, thus improving relations between stakeholders.

4.4.2 Combining Smart Contracts to Create a Decentralized Autonomous Organization

A group of smart contracts can be used to create a decentralized autonomous organization (DAO). This is an organization that is executed by rules, which are encoded as computer programs, using smart contracts. IoT devices and the large number of measurements and monitors that can be placed in buildings are a good reason for a building to be created as a DAO at the beginning of a project, during the construction phase and until the use phase.

The integration of blockchain with the building maintenance system (BMS) can lead the DAO of the building to order a new installation of light, to accept delivery and responsibility for it, to call someone to install it and to pay the supplier and the installer. The payment can be made from the DAO wallet, which is linked to the wallets of those living in the building. Rents can be collected and corporate fees and insurance payments can be managed independently by the DAO of the building.

The construction phase is not so different. It just needs more human input to clarify the requirements and make decisions to meet them – for example, which light installation, room temperature range, etc. All requirements and decisions start a series of *if–then* conditions. They use packages of interconnected smart contracts executed between a customer and various members of the project team, a general contractor, a subcontractor to design, monitor, approve, conduct tenders, install, certify and deliver the constructed asset.

Project management can also be included in blockchain – records of approvals at the pre-construction phase, but also during the use of the building, for voting on different issues requiring approval.

4.4.3 Building Information Modeling

The construction sector combines large teams that shape and design the built environment. The advent of technology, especially IoT and BIM, across the sector leads to more opportunities for collaboration and the emergence of new ideas. This power can be used to highlight the use of blockchain technology. BIM is a computer model that contains a wide range of asset information such as 3D geometry, construction management information (for example, time schedules and costs) or performance and maintenance metrics. The combination between BIM and blockchain technology can serve as a single source of truth.

BIM technology serves as a single source of truth for data. Placing the data validation, design approvals and project management solutions on a blockchain can result in a unified source of truth that includes all aspects of the project. BIM can combine blockchain information such as supply chain information, origin of materials, payment data, etc., especially during construction. It can also assign information to the blockchain such as design solutions, data source or orders for change of a model. Later, this information can be used by smart contracts to initiate further actions such as payments or orders of materials.

Blockchain, IoT and BIM are foundations that lead to smart asset management, because the projects do not stop at the asset delivery, but are transformed and continue until the end of its life cycle.

The combination between BIM and blockchain can significantly increase the smart contract's efficiency. For example, a BIM model can be used as part of a contract between participating partners. Therefore, all of them work to match the real physical project construction with the BIM model in the contract. If there is any deviation from the model, it may lead to change or rework of orders. The inclusion of BIM in smart contracts can also authorize payment only when the project construction is built in accordance with the digital plans.

Combining BIM and blockchain can also work to retain all project parties and create a higher level of transparency. The combination can actually improve the efficiency of BIM technology. BIM currently uses partner networks to share information, but blockchain can make real-time updates. This constant feedback and monitoring would help transparency and lead to better overall communication and higher-quality project workmanship.

Combining BIM and blockchain technology can also retain all parties in the project and create a higher transparency level. Actually, this combination can improve the efficiency of BIM technology. Currently, BIM shares information through partner networks, but the blockchain can make updates in real time. These constant monitoring and feedback help transparency and lead to better overall communication and higher-quality workmanship of the project.

4.4.4 Decentralized Network Management of Devices

In order for a blockchain to be successfully implemented, it needs to work with "digitally native" assets – assets that can be successfully presented in digital format. IoT and other tools allow the digitization of non-digital assets. Having in mind the wide application of IoT devices in the construction sector, blockchain technology can be considered as the backbone of a decentralized network of IoT devices. Thus, the blockchain can serve as a public ledger for a huge number of devices that will no longer need a central hub for mediation in communication between them.

4.4.5 Rationalization of Financing and Payments

The blockchain concept is built around exchange of money. It can be used for rationalization of payment processing. Many companies face significant challenges in ensuring payments on time to all stakeholders in a project.

A common practice for a contractor in construction projects is to request an advance payment to cover significant start-up or public procurement costs that need to be incurred before construction begins. In these cases, the customer should request an advance payment guarantee to secure the payment against default by the contractor.

Blockchain technology can significantly reduce paperwork from more than several days to less than 24 hours.

Issues related to late payments and cash flow are persistent problems in the construction sector. One of the most relevant blockchain applications in the construction industry is embedding a blockchain-based platform into the project implementation practice. It can initiate payments based on digitally approved work, smart contract actions and contract terms.

Blockchain technology can improve the financing and payment processes by increasing security and creating traceable information. The platforms, which are designed to change how subcontractors are hired and paid, are a very good example of technologies that play an important role in subcontracting relationships. They can provide faster payments and promote better relationships between partners.

4.4.6 Compliance Simplification

Providing evidence of compliance (for example, taxes, registrations, etc.) is often difficult, expensive and time consuming. Placing information related to evidence in the blockchain can significantly reduce the administrative burden and costs. Also, it can generally help to increase the availability of information during the bidding and implementation of a project. In addition, linking blockchain technology to emerging technologies such as artificial intelligence and IoT can enable collecting and processing of data in real time to improve overall compliance. The project information registered in the blockchain can be easily used to prove compliance with regulatory requirements.

4.4.7 Supply Chain Management (Origin and Traceability)

Blockchain technology has an important role in supply chain management. In particular, a blockchain can help in tracking physical elements and materials from their place of origin to their final destination. It can improve transparency, which in turn can help all partners to understand and agree with each other, and avoid potential difficulties and failures.

Issues related to sustainability make customers demand proof of the products' legitimacy and authenticity. Using the supply chain to track goods and assets is important for improvement of operational impact. Knowing the full journey of a product provides several advantages such as improved product safety, reduced fraud and increased accuracy in forecasting and joint planning in the sector. If the supply chain of a company breaks or delivers a defective final product, it has a financial impact on the company.

The origin of the assets is a critical aspect of quality assurance and control. It is achieved through data extracted from IoT devices, various vendors and their systems using application programming interface (API) integrations. The origin of construction materials and the creation of a verified custody chain ensure transparency for all partners in the supply chain and improve the assessment of the construction sector for sustainability. Reliable suppliers which have high-quality products can be recognized and encouraged to maintain quality certificates. In this way long-term relationships can be established.

4.4.8 3D Printing of New Construction Parts

3D printing and additive manufacturing (building 3D objects by adding materials layer upon layer) are processes which are highly driven by technologies. The digital files included in them can be easily transferred with one click. Therefore, parts and products are easier to track and share. This leads to the creation of smarter digital supply networks and chains. Using blockchain to support these evolving infrastructures can optimize project management, eliminate security vulnerabilities and protect intellectual property from theft. Ultimately, it can help the 3D printing and additive manufacturing sectors to develop and scale.

4.5 CONCLUSION

Construction is a highly regulated sector, employing a wide variety of partners in complex projects. Validation of their identity, quality of work and reliability can be a difficult and time-consuming task. This challenge can be addressed by an ecosystem, based on blockchain technology, which can make it easier for general contractors to track progress and verify identities across multiple teams. Blockchain technology can also ensure that construction materials are delivered from the right places and have the appropriate quality. Smart contracts can facilitate the automatic payment of timely payments related to the milestones of the project.

Blockchain technology has tremendous potential to be a highly positive force for change in the construction sector. Although the sector is not yet ready for its full implementation, the adoption of such technology is closer to reality than far away. The full value of the blockchain can be realized through the cooperation of different parties in the construction. Therefore it is beneficial to share best practices and experience in its implementation through partnerships and joint pilot projects. Looking to the future, it is only a matter of time before blockchain is necessary in any construction business.

Due to many research and practical works, indicating ongoing digitalization projects, and the clear change in how the construction sector operates, the author considers that there is a place for blockchain smart contract application in the future. Also, a strong opportunity to incorporate blockchain technology along with BIM technology has been identified. These technologies have become more common in the construction sector. In addition, willingness to share information between all participants shows a solid basis for the sector to use a blockchain solution. Blockchain can address many of the current concerns in the sector such as data security, privacy and speed of transaction, and provide better construction project management, ensuring a more efficient experience for all participating professionals and clients.

A lot of research still needs to be done in the field of blockchain technology and its implementation. In addition, studies should indicate and describe the success level of the implementation in terms of identifying the actual efficiency benefits of the technology usage. Further research should also be carried out on the various blockchain configurations, which try to establish a certain level of consensus in use cases and definitions from the construction sector. Another interesting aspect for further research is the potential of blockchain and smart contracts to improve construction supply chains in terms of sustainability, as the construction sector plays a huge role in the production of greenhouse gas emissions.

REFERENCES

Ablyazov, T., & Petrov, I. (2019). Influence of blockchain on development of interaction system of investment and construction activity participants. IOP Conference Series: Materials Science and Engineering, Volume 497, 012001. https://doi.org/10.1088/1757-899X/497/1/012001

Andoni, M., Robu, V., Flynn, D., Abram, S., Greach, D., Jenkins, D., McCallum, P., & Peacock, A. (2019). Blockchain technology in the energy sector: A systematic review of challenges and opportunities. Renewable and Sustainable Energy Reviews, Volume 100, pp. 143–174. https://doi.org/10.1016/j.rser.2018.10.014

Brakeville, S., & Perepa, B. (2019). Blockchain Basics: Introduction to Distributed Ledgers. IBM. Available at: https://developer.ibm.com/technologies/blockchain/tutorials/cl-blockchain-basics-intro-bluemix-trs/ (accessed 20.08.2020).

Business Blockchain HQ. Blockchain Fundamentals. Available at: https://businessblockchainhq.com/blockchain-fundamentals/ (accessed 21.08.2020).

Cong, L., & He, Z. (2019). Blockchain disruption and smart contracts. The Review of Financial Studies, Volume 32, Issue 5, pp. 1754–1797. https://doi.org/10.1093/rfs/hhz007

Coyne. R., & Onabolu, T. (2018). Blockchain for architects: Challenges from the shared economy. Architectural Research Quarterly, Volume 21, Issue 4, pp. 369–374. https://doi.org/10.1017/S1359135518000167

Crosby, M., Nachiappan, Pattanayak, P., Verma, S., & Kalyanaraman, V. (2015). Blockchain Technology: Beyond Bitcoin. Sutardja Center for Entrepreneurship & Technology Technical Report, University of California, Berkeley.

Feig, E. (2018). A Framework for Blockchain-Based Applications. Available at: https://arxiv.org/pdf/1803.00892.pdf (accessed 20.08.2020).

Gupta, M. (2018). Blockchain for Dummies. John Wiley, Hoboken, NJ.

Hammi, A., & Bouras, A. (2018). Towards safe-BIM curricula based on the integration of cybersecurity and blockchain features. In: Proceedings of INTED2018 Conference, Valencia, Spain, pp. 2380–2388.

Hughes, D. (2017). The Impact of Blockchain Technology on the Construction Industry, Medium. Available at: https://medium.com/the-basics-of-blockchain/the-impact-of-blockchain-technology-on-the-construction-industry-85ab78c4aba6 (accessed 26.08.2020).

Kamble, S., Angappa, G., & Arha, H. (2018). Understanding the blockchain technology adoption in supply chains – Indian context. International Journal of Production Research, Volume 57, Issue 7, pp. 2009–2033.

Lanko, A., Vatin, N., & Kaklauskas, A. (2018). Application of RFID combined with blockchain technology in logistics of construction materials. MATEC Web of Conferences, Volume 170, Issue 03032.

Li, J., Greenwood, D., & Kassem, M. (2018). Blockchain in the construction sector: A socio-technical system framework for the construction industry. In *Advances in Informatics and Computing in Civil and Construction Engineering: Proceedings of the CIB W78 2018*. Springer.

Liu, Y., Van Nederveen, S., & Hertogh, M. (2017). Understanding effects of BIM on collaborative design and construction: An empirical study in China. International Journal of Project Management, Volume 35, Issue 4, pp. 686–698.

Mason, J. (2017). Intelligent contracts and the construction industry. Journal of Legal Affairs and Dispute Resolution in Engineering and Construction, Volume 9, Issue 3.

Mason, J., & Escott, E. (2018). Smart Contracts in Construction: Views and Perceptions of Stakeholders. University of the West of England. Available at: http://eprints.uwe.ac.uk/35123/3/FIGrevised.pdf (accessed 26.08.2020).

Mathews, M., Robles, D., & Bowe, B. (2017). BIM+blockchain: A solution to the trust problem in collaboration? In: CITA BIM Gathering, 2017.

Maxwell, W., & Salmon, J. (2017). A Guide to Blockchain and Data Collection. Hogan Lovells. Available at: www.hlengage.com/_uploads/downloads/5425Guidetoblock chainV9FORWEB.pdf (accessed 21.08.2020).

Nakamoto, S. (2008). Bitcoin: A Peer-to-Peer Electronic Cash System. Bitcoin. Available at: https://bitcoin.org/bitcoin.pdf (accessed 20.08.2020).

Nofer, M., Gomber, P., Hinz, O., & Schiereck, D. (2017). Blockchain. Business & Information Systems Engineering, Volume 59, Issue 3, pp. 183–187.

Penzes, B. (2018). Blockchain Technology in Construction Industry. Institution of Civil Engineers. Available at: www.ice.org.uk/ICEDevelopmentWebPortal/media/Documents/News/Blog/Blockchain-technology-in-Construction-2018-12-17.pdf (accessed 20.08.2020).

Puthal, D., Malik, N., Mohanty, S.P., Kougianos, E., & Yang, C. (2018). The blockchain as a decentralized security framework [future directions]. IEEE Consumer Electronics Magazine, Volume 7, Issue 2, pp. 18–21. Available at: www.researchgate.net/publication/323491592_The_Blockchain_as_a_Decentralized_Security_Framework_Future_Directions (accessed 20.08.2020).

Swan, M. (2015). Blockchain: Blueprint for a New Economy. O'Reilly Media, Sebastopol, GA.

Swan, M. (2017). Anticipating the economic benefits of blockchain. Technology Innovation Management Review, Volume 7, pp. 6–13.

Tapscott, D., & Tapscott, A. (2017). How blockchain will change organizations. MIT Sloan Management Review. Available at: https://sloanreview.mit.edu/article/how-blockchain-will-change-organizations/ (accessed 20.08.2020).

Wang, J., Wu, P., Wang, X., & Shou, W. (2017). The outlook of blockchain technology for construction engineering management. Frontiers in Engineering Management, 2017, Volume 4, Issue 1, pp. 67–75. https://doi.org/10.15302/J-FEM-2017006

Weber, I., Lu, Q., Tran, A., Deshmukh, A., Gorski, M., & Strazds, M. (2019). A Platform Architecture for Multi-Tenant Blockchain-Based Systems. International Conference on Software Architecture, Hamburg, Germany. http://hdl.handle.net/102.100.100/86005?index=1

Zheng, Z., Xie, S., Dai, H., Chen, X., & Wang, H. (2017). An overview of blockchain technology: Architecture, consensus, and future trends. In: IEEE 6th International Congress on Big Data, 2017, pp. 557–657.

5 Artificial Neural Network Applications in Social Media Activities

Impact on Depression during the COVID-19 Pandemic

Marta R. Jabłońska

5.1 INTRODUCTION

Coronavirus disease (COVID-19) is a declared global pandemic threatening global health (Chan, Nickson, Rudolph, Lee and Joynt 2020; Kakodkar, Kaka and Baig 2020). The World Health Organization (WHO) announced the virus severe acute respiratory syndrome coronavirus 2 (SARS-CoV-2) as a public health emergency of international concern; as a result of global logarithmic expansion of COVID-19, WHO declared it to be a pandemic that was deeply concerning by the alarming levels of spread, severity, and inaction on the 11 March 2020 (Boulos and Geraghty 2020; Kakodkar, Kaka and Baig 2020; Miri, Roozbeh, Omranirad and Alavian 2020; Nature 2020; WHO 2020a,b). On the day of writing this chapter 32,454,698 people had been diagnosed with COVID-19 and 988,496 had died (Worldometer 2020).

All around the world, governments have been responding to this pandemic with steps previously taken only in times of war (*Nature* 2020). In Poland, on 4 March 2020, the Minister of Health informed about the first case of COVID-19 in the country. Five days later, the government decided to introduce new actions to protect

Email: marta.jablonska@uni.lodz.pl

DOI: 10.1201/9781003216278-5

Polish society (Website of the Republic of Poland 2020) (Table 5.1). According to the Prime Minister, Mateusz Morawiecki, Poland was "one of the first, if not the first country in Europe, which have made such decisions." In a short period, life in Poland had drastically changed, with the majority of citizens staying at home. This state continued until May 2020, when restrictions began to be eased; many of them still apply today. The period between 12 March and the end of May 2020 was called "lockdown" and affected the lives of all Poles, challenging them also in psychological terms.

TABLE 5.1
Actions Taken by the Polish Government against the Spread of COVID-19

Date	Regulations
9 March 2020	Sanitary controls were brought in at the main borders with Germany and the Czech Republic, expanding the scope of these controls to other border crossings. Controls implemented at all cars, rail, and port crossings so as not to disturb the flow of traffic at the border
12 March 2020	A state of epidemic threat was declared. The functioning of education system units was temporarily suspended
13 March 2020	Galleries and shopping malls (excluding pharmacies, grocery stores, laundries, banks, financial institutions, and drugstores) were closed; indoor and outdoor gatherings of over 50 people of public, state, and religious nature were banned
15 March 2020	The Polish borders were closed to foreigners. All Polish citizens who were abroad could return with a mandatory 14-day home quarantine. All international passenger air and rail connections were suspended
23 March 2020	Quarantine was tightened: free movement was banned except for living, health, and professional purposes, limiting the number of people using public transport, prohibition of assemblies, and limiting to five the number of people allowed to participate in religious rites at the same time
1 April 2020	The number of customers in stores, markets, and post offices were limited (three people were allowed at each cash or paying position); large-scale construction stores were closed on weekends; public institutions were required to perform their duties remotely; disposable gloves were required in shops; stores were open exclusively to seniors between 10:00 and 12:00; hotels, hairdressers, cosmetics, tattoo, piercing salons, rehabilitation, and massage salons (including home visits) were closed. Children and adolescents under 18 years were not permitted to leave the home unattended. It was forbidden to use parks, beaches, boulevards, promenades, and city bikes. The obligation to maintain a distance of at least 2 meters between pedestrians was introduced; this also applied to families and loved ones
16 April 2020	A new obligation was introduced that the mouth and nose should be covered. School and higher education exams were postponed

As much as we want to comprehend the psychological results of COVID-19, we are faced with a hindrance to this understanding due to the vast amount of misinformation created by a limited amount of research in this particular area. COVID-19 may affect an emotional state of being, causing symptoms of anxiety or depression, and social media may fuel them (Wiederhold 2020). WHO has even provided a new term for social media and COVID-19 information, called "infodemic," an overabundance of information – some accurate and some not – that makes it hard for people to find trustworthy sources and reliable guidance when they need it (WHO 2020c). Some researchers have already studied this relation (Iacobucci 2020; Lohiniva, Sane, Sibenberg, Puumalainen and Salminen 2020; Taheri, Falahati, Radpour, Karimi, Sedaghat and Karimi 2020). These findings raise a disturbing issue in light of the global COVID-19 epidemic in relation to the impact of social media on overall well-being. However, several studies have concluded that social media has had a positive impact on communications on COVID-19 (Chan, Nickson, Rudolph, Lee and Joynt 2020; Iacobucci 2020). Yet, this research has also pointed out the risks associated with social media, such as biased knowledge dissemination, mixing real and false information, and alleviating COVID-19 patients' fears (Llewellyn 2020).

Artificial neural networks (ANNs) have recently been implemented in psychological studies as they focus on a problem at a relatively abstract level (Chance et al. 2020). These applications included a variety of topics, i.e., IQ assessment, emotion recognition (Tu et al. 2020), post-traumatic stress disorder screening (Kaczmarek et al. 2019), or psycho-diagnostics data of youth analysis (Slavutskaya et al. 2019). Although researchers used ANN for COVID-19-related studies, currently no research is known on the impact of the virus on depression levels assessed through these tools.

5.2 METHODS

5.2.1 Data Collection

This study focused on the impact of reading COVID-19 content on social media on depression. A total of 871 respondents from Poland filled in a questionnaire during lockdown. After records with missing data were removed, 761 were valid. Due to difficulties in collecting data caused by the pandemic and lockdown, no statistical methods were used to predetermine sample size and selection. Respondents were chosen with a snowball method. Since the purpose of the study took into account social media usage, an invitation to the study was announced in these media (Facebook, Instagram) with a request to disseminate it further. The restriction in the selection of the sample was the place of residence (Poland) and being an adult, verified with appropriate questions. Data were collected from 1 April to 30 May.

5.2.2 Data Analysis

The first part of the questionnaire started with questions related to COVID-19 and social media activities during lockdown: number of virtual contacts, reading and publishing COVID-19 content, using particular social media, feelings after such reading, and performing activities related to subjective improvement in well-being (hobby, meditation, yoga). Next, respondents were asked to fill in scales self-measuring life satisfaction, anxiety, and depression. Satisfaction with life was assessed with a scale developed by Diener, Emmons, Larsen, and Griffin (1985). Satisfaction With Life Scale (SWLS) is a valid and reliable 5-item measure of the factors of subjective well-being, appropriate for use with a broad range of age groups and areas. It is recommended for emotional well-being assessment due to respondents' conscious evaluative judgment (Pavot and Diener 1993). In this study, Cronbach α for the life satisfaction scale was 0.892. The Hospital Anxiety and Depression Scale (HADS) is a useful self-report questionnaire for anxious and depressive state screening in non-psychiatric settings (Annunziata, Muzzatti, Bidoli, Flaiban, Bomben, Piccinin, Gipponi, Mariutti, Busato and Mella 2020). It consists of two 7-item subscales, anxiety (HADS-A) and depression (HADS-D). A score exceeding 8 points for each subdomain indicates clinically significant symptoms (Slenders, Verberne Visser-Meily den Berg-Vos Kwa and van Heugten 2020). Here, Cronbach α was 0.791 and 0.801 for the HADS-A and HADS-B, respectively. Finally, participants were asked about age and gender.

Life satisfaction and anxiety levels were calculated based on the above scales as numeric variables. Depression was calculated as a numeric variable but also coded on an ordinal scale according to the HADS instruction. In this case, three levels were recognized: normal, borderline abnormal, and abnormal.

In the study, two ANN algorithms were used: multilayer perceptron (MLP) and radial basis function (RBF). Both are so-called feed-forward networks, due to one-directional information processing from input to output neurons. MLPs have three layers: an input, a more hidden, and an output layer (Kalogirou 2000). In this algorithm neurons are interconnected; each node receives the input data from other neurons and passes through the hidden to an output layer. MLPs used supervised learning where a set of input–output pairs was used for learning and modeling dependencies within these pairs. Learning starts with small random, initial weights that are adjusted during training in order to minimize error. Back-propagation is used to make those weigh and bias adjustments relative to the error, a one-pass back-propagation algorithm comprised of a forward pass and a backward pass. The first propagates the input vector through the network to adjust the weights so the output neuron is a result of weighted inputs. In the backward pass, the weights and biases are back-propagated through the network to give an information on error, to adjust the parameters aiming to minimize the error.

RBF networks are easy to design, tolerant to input noise, fast and comprehensive during learning, and they are able to respond to new patterns (Yu et al. 2011). They are often slower and larger than MLPs, and perform less well in models with a vast number of inputs. This is due to their high sensitivity to switch on unnecessary inputs. In Statistica, RBF networks are built similarly to MLP but they possess a single hidden layer and no weights (Fath et al. 2020).

All ANN implementations in this study were made in an ANN automatic Statistica module which allows the following parameters to be modified: structure (RBF or MLP), error function (sum of squares or mutual entropy), MLP activation and output neuron functions (linear, logistic, hyperbolic tangent, exponential, and sine), enabling/

disabling weight reduction, and the number of hidden and output neurons. Default stopping conditions in an automatic mode were unmodifiable. As initial weights and data assignment to the training, validation, and test sets were all randomly set, the results obtained may differ after re-performing calculations.

Dependency description methods are used to build models reflecting existing relationships between the set of input – explanatory variables – and the set of output – explained variables. When building regression neural networks, the dependent variable is quantitative, and the explanatory variables are quantitative or qualitative. Quantitative variables used in a network require initial preprocessing, which usually comes down to re-scaling them (MLP networks) or standard deviation (SD) method (RBFs). Re-scaling the values of the input variables adjusts their values to the range of values of the output neurons. This process is aimed at transforming the values of the input neurons to the interval for which the degree of variability of the logistic activation functions used in the neurons of the hidden and output layers is the highest. The use of the SD method leads uses zero–one standardization where the transformed sequence has a mean value of 0 and SD is equal to 1. In the case of input qualitative variables, pre-processing is a value conversion to a numerical form.

5.3 RESULTS

5.3.1 Regression Model

The aim of regression networks was to predict a numeric value of depression level based on the available variables described in the data analysis section. In these models, the results were quite satisfactory, as the best obtained network had a quality of testing of 77%. The MLP 15-35-1 network proved to be the most suitable. The numbers indicate the network topology. In this case it is an MLP with 15 neurons in the input layer, 35 in the hidden one, and one output neuron representing the predicted numerical depression value for the analyzed record. The network quality of the training, test, and validation sets was 0.734, 0.771%, and 0.653, respectively. This quality is expressed by the linear correlation coefficient between the values of the dependent variable and the outputs of the network. The learning algorithm in this network was the Broyden–Fletcher–Goldfarb–Shanno (BFGS) 21 and the number indicates the epochs needed for the learning process to converge. The error function was sum-of-squares (SOS), and the activation functions in the neurons of the input and hidden layers were exponential and hyperbolic tangent, respectively.

In order to graphically present the network results, a graph was made showing the values of the dependent variable in relation to the network predictions (Figure 5.1). On analysis, a relatively good fit of the network to the analyzed depression levels can be seen. The validity of the use of variables in the network was checked by means of a global sensitivity analysis. This shows how the quality of the model will change after one of the variables has been excluded. Values above 1 indicate that the model would deteriorate in quality after removing a given variable, and values below 1 that the quality could improve. The results in Table 5.2 show that the use of all the variables in the network was justified.

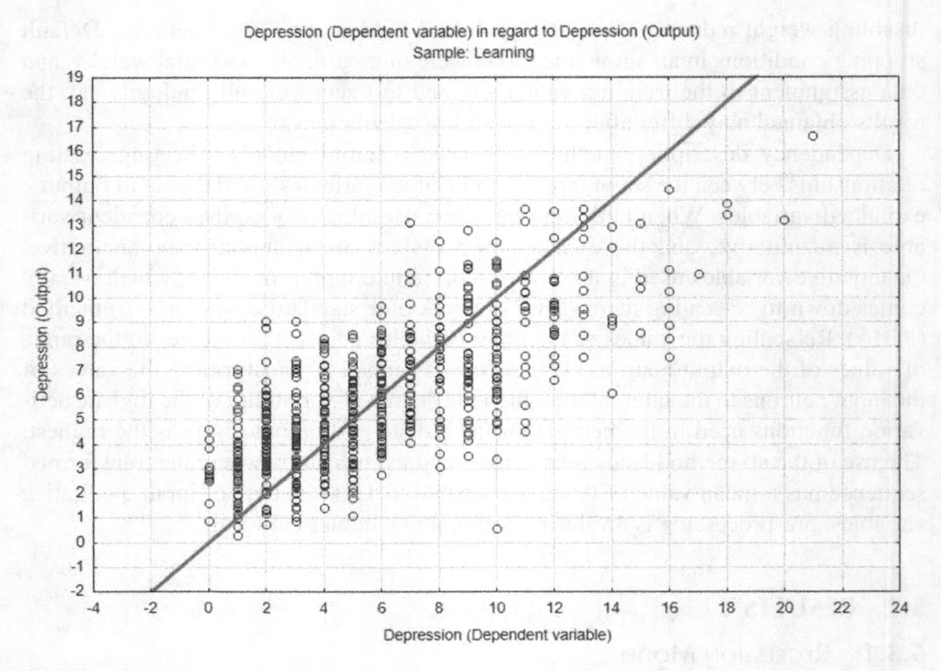

FIGURE 5.1 Graph of the fit of the network to the predicted depression level.

TABLE 5.2
Global Sensitivity Analysis Results for the Multilayer Perceptron (MLP) 15-35-1 Network

Variable	Parameter
Anxiety	1,650
Life satisfaction	1,086
Reading COVID-19 content	1,083
Hobby	1,050
Publishing COVID-19 content	1,027
Yoga	1,021
Fear after reading	1,018
YouTube	1,018
Age	1,015
Instagram	1,008
Facebook	1,007
Twitter	1,007
Virtual contacts	1,004
Meditation	1,002
Female	1,002

TABLE 5.3
Weight Values Assigned to Specific Connections in the Multilayer Perceptron (MLP) 15-35-1 Network

Connections	Weights
Hidden neuron 21 → Depression	0.496753
Hidden neuron 5 → Depression	0.489574
Anxiety → Hidden neuron 5	0.444804
Hidden neuron 12 → Depression	0.397166
Hidden neuron 17 → Depression	0.385960
Anxiety → Hidden neuron 4	0.312685
Hidden neuron 33 → Depression	0.295733
Anxiety → Hidden neuron 35	0.293862
Anxiety → Hidden neuron 15	0.279612
Hidden neuron 8 → Depression	0.267935
Anxiety → Hidden neuron 21	0.258224
Hidden neuron 28 → Depression	0.251742
Hidden neuron 32 → Depression	0.234994
Hidden neuron 11 → Depression	0.234399
Hidden neuron 4 → Depression	0.222659
Anxiety → Hidden neuron 24	0.209395
Anxiety → Hidden neuron 29	0.200725
Hidden neuron 14 → Depression	0.174304
Anxiety → Hidden neuron 1	0.173586
Anxiety → Hidden neuron 33	0.173045

Note: Due to the large connections set ($N = 596$), only the first 20 results are presented.

Table 5.3 presents the weight values assigned to specific connections in the network. They do not have a substantive interpretation, but small values may suggest a negligible impact of a given connection on the network response.

The next step was to test the network's ability to predict new data. For this purpose, a new record, not from the original database, was introduced into the network. After calculating the value of the depression level variable on the basis of the questionnaire data, the result was 3. After entering the data into the network, the network predicted a value of 2.91. Taking into account that the HADS test, on the basis of which the level of depression was estimated, does not use fractional values in its scale, the result would be rounded to 3 and therefore this would be equal to the actual value calculated from the respondent's answers.

5.3.2 Classification Models

The purpose of the classification model is to assign the tested record to one of the known classes. In the case analyzed, it is a level of depression, recoded from quantitative to ordinal scale, in accordance with the HADS instruction. As a result, each

TABLE 5.4
Error Matrix for Multilayer Perceptron (MLP) 15-127-3 Network

	Depression			
	Abnormal	**Borderline abnormal**	**Normal**	**All**
Total	67,000	103,000	363,000	533,000
Correct	42,000	7,000	353,000	402,000
Incorrect	25,000	96,000	10,000	131,000
Correct (%)	62,687	6,796	97,245	75,422
Incorrect (%)	37,313	93,204	2,755	24,578

respondent was assigned to one of three classes, ordered as follows: normal, borderline abnormal, and abnormal. Therefore, in the classification model, the output variable is nominal, and the input variables can be quantitative and qualitative. The best-fitting model was MLP 15-127-3 with a 75.44% quality of testing. Just like in the regression model, the numbers indicate the network topology: 15 input, 127 hidden, and three output neurons. The last value is based on the number of classification options (normal, borderline abnormal, and abnormal). The network quality of the training and validation was 75.42% and 60.53% accordingly. The learning algorithm in this network was the BFGS with nine epochs. The error function was SOS, and the activation functions in the input and hidden neurons were both hyperbolic tangent.

Classifying networks can be assessed using a matrix of errors, specifying the number of correctly and incorrectly classified cases (Table 5.4).

Although the network recognized 75.42% of cases correctly, its performance for the borderline abnormal class was poor as the network misclassified 93.2% of the cases belonging to this group. The assessment of the network is additionally illustrated by the lift and gain charts for each class (Figures 5.2–5.4).

The gain chart is a graphical summary of the usefulness of models to predict the value of the categorized dependent variable. It illustrates what percentage of cases belonging to the class was comprised of the subsets of data containing the fractions of cases with the highest probability of being in a given class resulting from the model. As a reference, a line is drawn representing a random selection of the corresponding number of cases from the data set. Instead, the lift chart shows how much more often in relation to the entire data set cases belonging to the studied class occur more often in the subsets of data containing the fractions of cases with the highest probability of being in this class, resulting from the model.

The poor model results in classifying the borderline abnormal class caused a second attempt to model construction. For this purpose, the depression level was recoded again. The levels indicating the occurrence of this phenomenon have been aggregated into one class (abnormal). In this way, the network will classify cases into people suspected of having this disorder or not (two-class classification). This approach resulted in a network with higher testing quality of 82.46%. The network MLP 15-46-2 was characterized by a significant reduction in the number of neurons

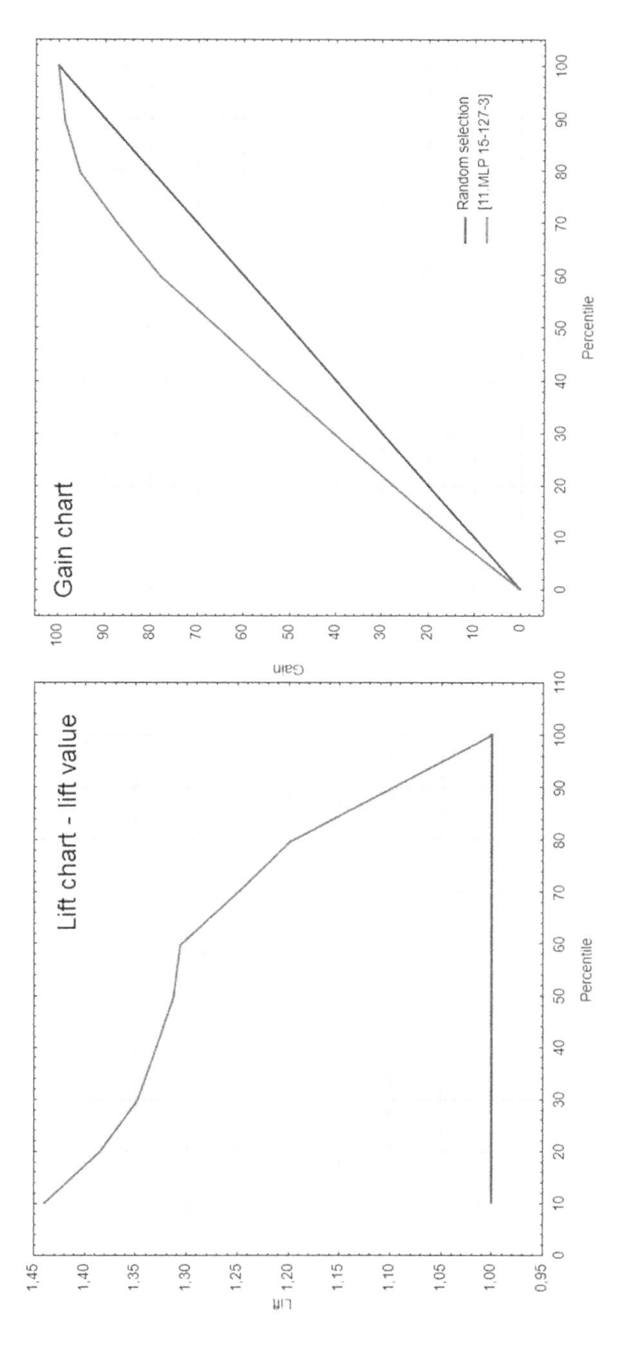

FIGURE 5.2 The lift and gain charts for a normal depression level class.

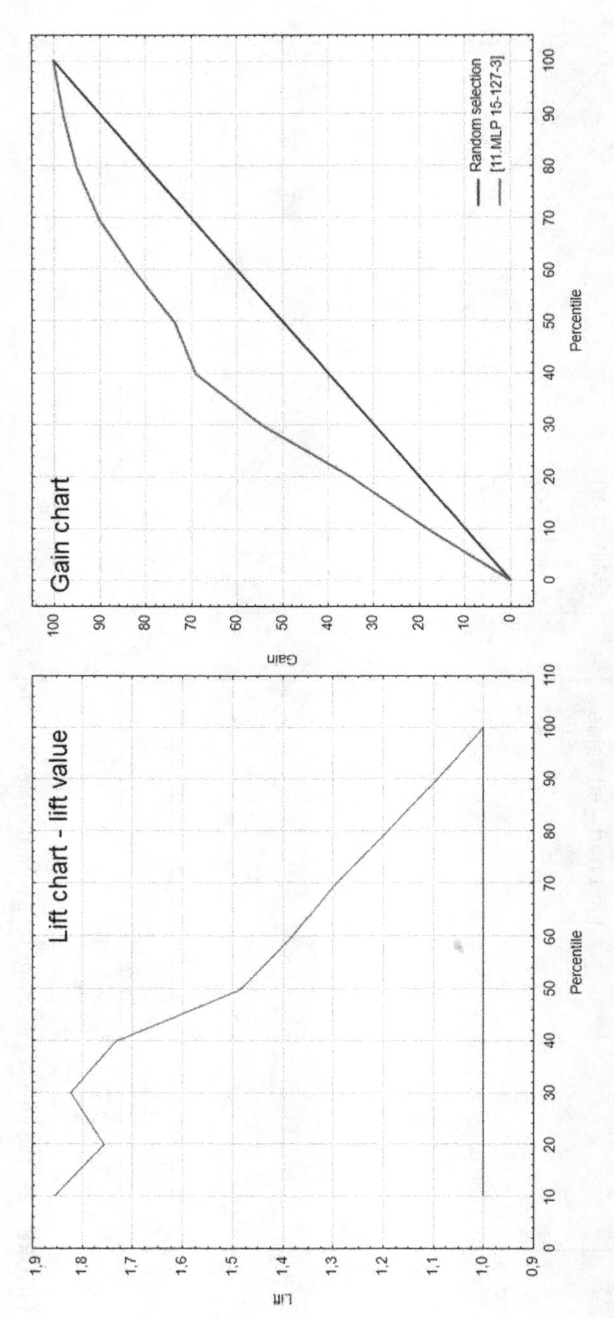

FIGURE 5.3 The lift and gain charts for a borderline abnormal depression level class.

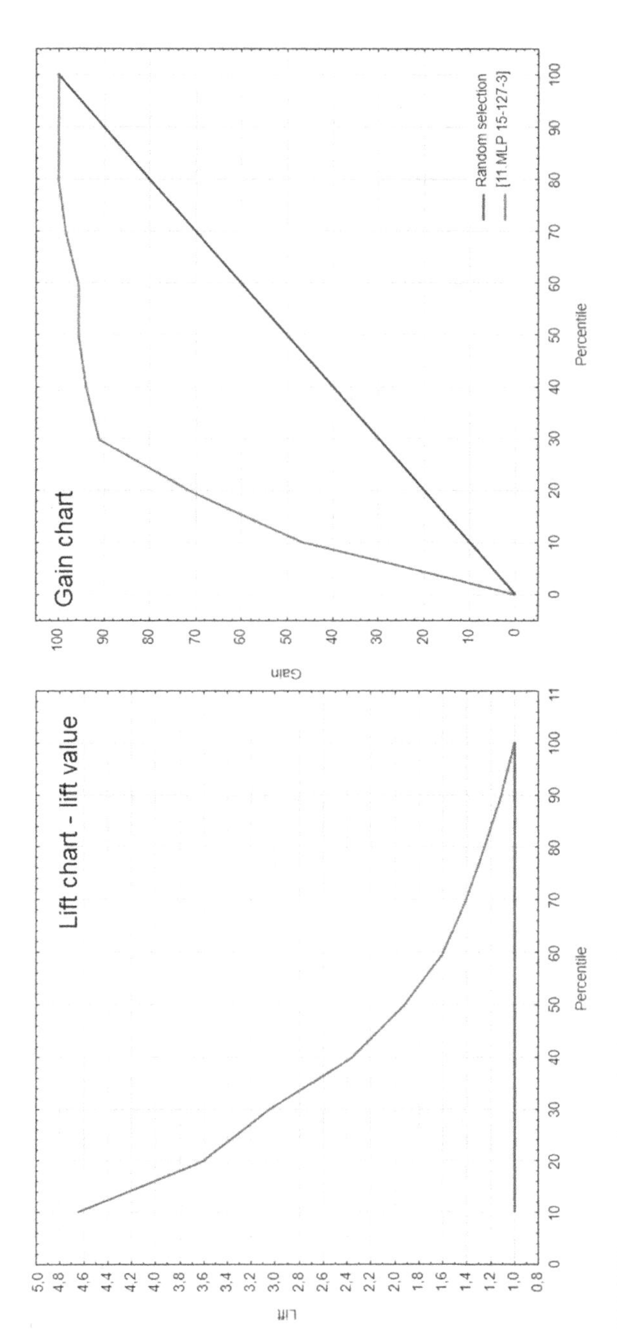

FIGURE 5.4 The lift and gain charts for an abnormal depression level class.

TABLE 5.5
Error Matrix for the Multilayer Perceptron (MLP) 15-46-2 Network

	Depression recoded		
	Abnormal	**Normal**	**All**
Total	170,000	363,000	533,000
Correct	101,000	344,000	445,000
Incorrect	69,000	19,000	88,000
Correct (%)	59,412	94,766	83,490
Incorrect (%)	40,588	5,234	16,510

in the hidden layers. The change in the number of output neurons from three to two was caused by a two-class classification (normal, abnormal). The quality of training was 83.49% and 75.44% for validation. The learning algorithm was the BFGS but the number of epochs increased to 20. The error function was SOS, and the activation functions in the input and hidden neurons were both hyperbolic tangent and exponential, accordingly. Table 5.5 presents the error matrix for this network and Figures 5.5 and 5.6 show lift and gain charts for normal and abnormal classes.

The results show that the second classification network model MLP 15-46-2 achieved better results. The proportion of correctly classified cases has increased. The lift and gain charts, compared to the first MLP 15-127-3 network, have been smoothed for the aggregated abnormal class, representing more averaged charts. In the models classifying objects into two classes, a tool useful for their evaluation is the receiver operating characteristic (ROC) curve. The area under the ROC curve is taken as a measure of the quality of the model. For the ideal network its value will be equal to 1. As the quality of the model deteriorates, the size of the area under the curve also decreases. Values equal to or less than 0.5 indicate low network usability. In this case the ROC curve shows good network operation (Figure 5.7).

5.4 DISCUSSION

Predicting and explaining human behavior is a classical problem of psychology. Psychological research aims to investigate and define the mechanisms, personality traits, and emotions that give rise to behaviors. Understanding and exploring human behavior may be supported by ANNs. By analyzing self-reported, medical, or other data these networks may help to discover new patterns that govern the way people behave.

The conducted research has shown that social media may impact mental health when it comes to reading COVID-19 content. Good network performance shows the relations between online actions related to COVID-19 content and depression levels. All presented models were able to predict with accuracy exceeding 70%. The new, previously unknown situation created by the coronavirus pandemic has created threats not only to life and physical health but also to mental health. The impact of the pandemic on mental health can be as dangerous in the long term as the effects of

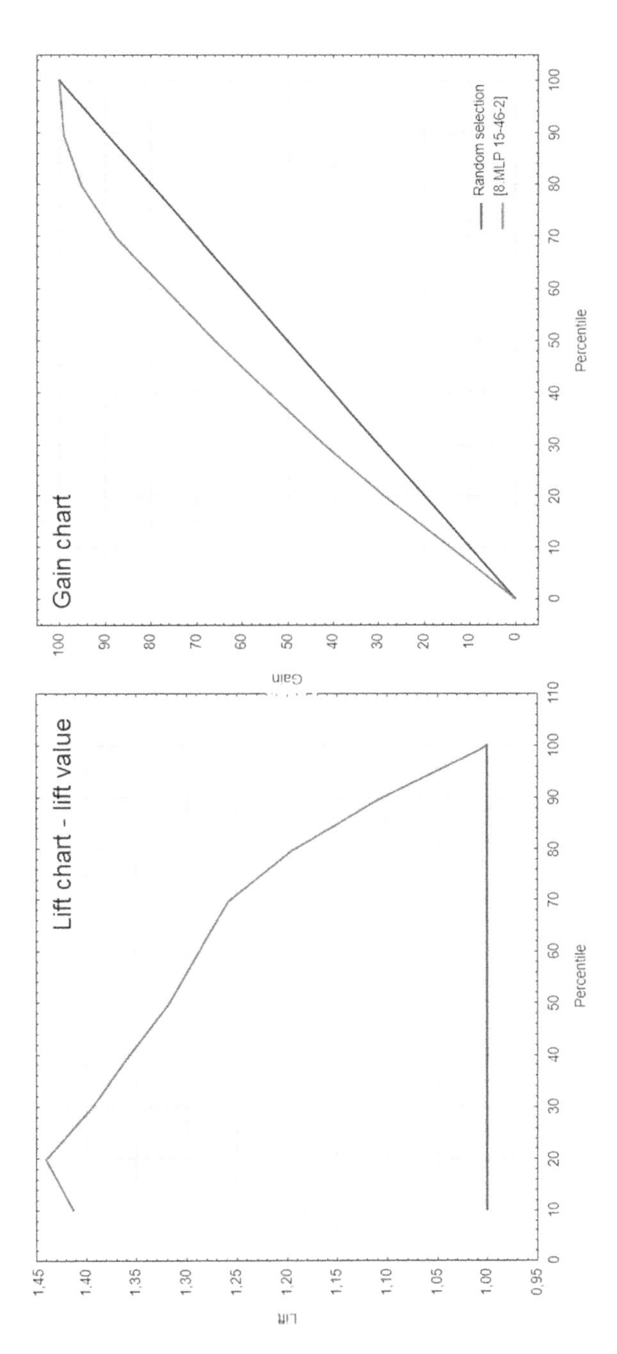

FIGURE 5.5 The lift and gain charts for a normal depression level class.

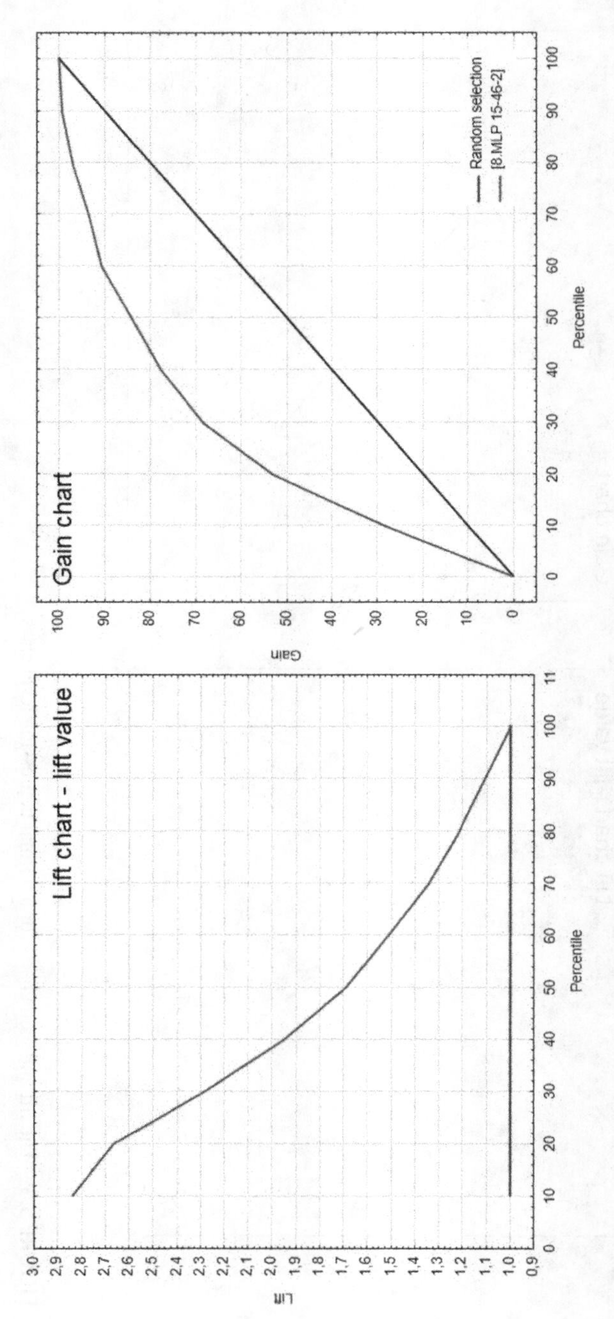

FIGURE 5.6 The lift and gain charts for a recoded abnormal depression level class.

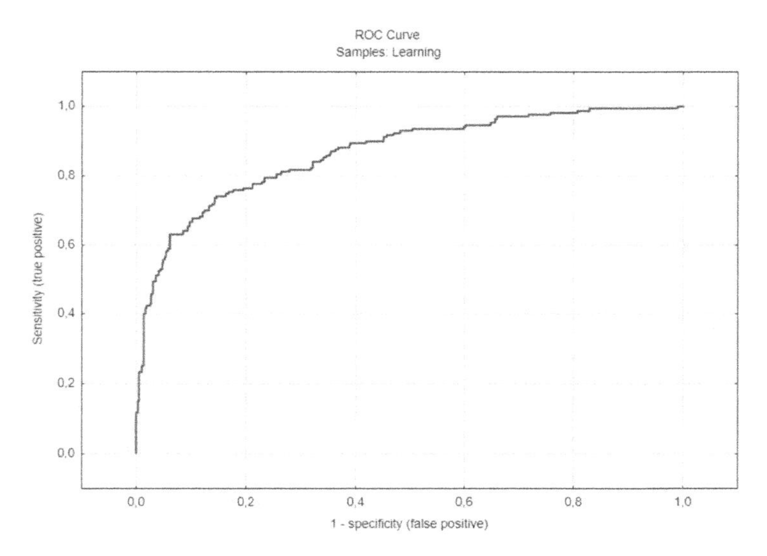

FIGURE 5.7 The receiver operating characteristic (ROC) curve for the multilayer perceptron (MLP) 15-46-2 network.

the disease itself. Moreover, the psychological effects are noted not only in infected patients, but also in healthy people. The pandemic sparked a massive publishing of social media content. Therefore, it is important to understand its effect on mental health.

This study focused solely on the impact of reading COVID-19 content on social media at the predicted level of depression. ANNs have proven to be effective tools for detecting this impact in regression and classification models. In both cases the MLP turned out to be the better topology, as they are usually faster and smaller than RBF networks.

Although the study seems to expand the current state of the art about the role of ANNs in psychological studies, it is not without certain limitations related to data collection. Each questionnaire-based study may be biased by participants providing distorted or preferential responses, as well as relying on their mental state on a self-measured basis. It should be stressed that self-assessment HADS are only valid for screening purposes; a definitive depression diagnosis must be conducted during a clinical examination. All data were gathered from Polish respondents as Poland was one of the first countries to introduce a lockdown. This is a limitation of the study as international research could show other relationships. Nevertheless, introducing new data from a different area to the constructed ANNs may be made without losing network performance.

As the pandemic is still ongoing and the effects of COVID-19 on mental health are not yet thoroughly recognized, this study contributes to a better understanding of this phenomenon and the role of ANNs in psychological studies.

BIBLIOGRAPHY

Anoor, M.M., Jahidin, A.H., Arof, H., Megat, A., Megat, S.A. (2020). EEG-based intelligent system for cognitive behavior classification. *J. Intell. Fuzzy Syst., 39(1),* 177–194. 10.3233/JIFS-190955.

Annunziata, M.A., Muzzatti, B., Bidoli, E., Flaiban, C., Bomben, F., Piccinin, M., Gipponi, K.M., Mariutti, G., Busato, S., Mella, S. (2020). Hospital Anxiety and Depression Scale (HADS) accuracy in cancer patients. *Support Care Cancer, 28,* 3921–3926. 10.1007/s00520-019-05244-8

Arpaci, I., Karataş, K., Baloğlu, M. (2020). The development and initial tests for the psychometric properties of the COVID-19 Phobia Scale (C19P-S). *Pers. Individ. Differ.,* 164, 110108. 10.1016/j.paid.2020.110108

Boulos, M.N.K., Geraghty, E.M. (2020). Geographical tracking and mapping of coronavirus disease COVID-19/severe acute respiratory syndrome coronavirus 2 (SARS-CoV-2) epidemic and associated events around the world: how 21st century GIS technologies are supporting the global fight against outbreaks and epidemics. *Int. J. Health Geogr., 19,* 8. 10.1186/s12942-020-00202-8

Burt, E.L., Atkinson, J. (2012). The relationship between quilting and wellbeing. *Am. J. Public Health, 34(1),* 54–59. 10.1093/pubmed/fdr041

Chan, A.K.M., Nickson, C.P., Rudolph, J.W., Lee, A., Joynt, G.M. (2020). Social media for rapid knowledge dissemination: early experience from the COVID-19 pandemic. *Anaesthesia,* 10.1111/anae.15057

Chance, F.S., Aimone, J.B., Musuvathy, S.S., Smith, M.R., Vineyard C.M., Wang, F. (2020). Crossing the cleft: Communication challenges between neuroscience and artificial intelligence. *Front. Comput. Neurosci., 14, 39.* 10.3389/fncom.2020.00039

Diener, E., Emmons, R.A., Larsen, R.J., Griffin, S. (1985). The Satisfaction with Life *Scale. J. Pers. Assess., 49,* 71–75.

Fath, A.H., Madanifar, F., Abbasi, M. (2020). Implementation of multilayer perceptron (MLP) and radial basis function (RBF) neural networks to predict solution gas-oil ratio of crude oil systems. *Petroleum, 6(1),* 80–91. 10.1016/j.petlm.2018.12.002

Iacobucci, G. (2020). Covid-19: diabetes clinicians set up social media account to help alleviate patients' fears. *BMJ, 368,* m1262. 10.1136/bmj.m1262

Kaczmarek, E., Salgo, A., Zafari, H., Kosowan, L., Singer, A., Zulkernine, F. (2019). Diagnosing PTSD using electronic medical records from Canadian primary care data. In Proceedings of the 6th International Conference on Networking, Systems and Security (NSysS '19). Association for Computing Machinery, New York, NY, USA, 23–29. 10.1145/3362966.3362982.

Kakodkar, P., Kaka, N., Baig, M. (2020). A comprehensive literature review on the clinical presentation, and management of the pandemic coronavirus disease 2019 (COVID-19). *Cureus, 12(4),* e7560. 10.7759/cureus.7560

Kalogirou, S.A. (2000). Applications of artificial neural-networks for energy systems. *Appl. Energy, 67(1–2),* 17–35. 10.1016/S0306-2619(00)00005-2.

Kelly, B. (2020). Coronavirus disease: challenges for psychiatry. *British Journal of Psychiatry 217(1),* 352–353. 10.1192/bjp.2020.86

Llewellyn, S. (2020). Covid-19: how to be careful with trust and expertise on social media. *BMJ, 368,* m1160. 10.1136/bmj.m1160

Lohiniva, A.L., Sane, J., Sibenberg, K., Puumalainen, T., Salminen, M. (2020). Understanding coronavirus disease (COVID-19) risk perceptions among the public to enhance risk communication efforts: a practical approach for outbreaks. *Euro. Surveill., 25(13),* 2000317. 10.2807/1560-7917.ES.2020.25.13.2000317

Miri, S.M., Roozbeh, F., Omranirad, A, Alavian, S.M. (2020). Panic of buying toilet papers: A historical memory or a horrible truth? Systematic review of gastrointestinal manifestations of COVID-19. *Hepat. Mon., 20(3),* e102729. 10.5812/hepatmon.102729

Nature (19 March 2020). COVID-19: what science advisers must do now. Nature, 579, 319–320. https://media.nature.com/original/magazine-assets/d41586-020-00772-4/d41586-020-00772-4.pdf: accessed 06 August 2020.

Pavot, W.G., Diener, E. (1993). Review of the Satisfaction with Life Scale. *Psychol. Assess., 5,* 164–172.

Slavutskaya, E.V., Abrukov, V.S., Slavutskii, L.A. (2019). Simple neuro network algorithms for evaluating latent links of younger adolescents' psychological characteristics. *Eksperimental'naâ psihologiâ = Exp. Psychol. (Russia), 12(2),* 131–144. 10.17759/exppsy.2019120210.

Slenders, J.P.L., Verberne, D.P.J., Visser-Meily, J.M.A., den Berg-Vos, R.M.V., Kwa, V.I.H., van Heugten, C.M. (2020). Early cognitive and emotional outcome after stroke is independent of discharge destination. *J. Neurol.,* 10.1007/s00415-020-09999-7

Taheri, M.S., Falahati, F., Radpour, A., Karimi, V., Sedaghat, A., Karimi M.A. (2020). Role of social media and telemedicine in diagnosis & management of COVID-19: An experience of the Iranian Society of Radiology. *Arch. Iran. Med. 23(4),* 285–286. 10.34172/aim.2020.15

Taylor, S., Landry, C.A., Paluszek, M.M., Fergus, T.A., McKay, D., Asmundson, G.J.G. (2020). COVID stress syndrome: concept, structure, and correlates. *Depress. Anxiety, 2020,* 1–9. 10.1002/da.23071

Tu, G., Fu, Y., Li, B., Gao, J., Jiang, Y., Xue, X. (2020). A multi-task neural approach for emotion attribution, classification, and summarization. *IEEE Trans. Multimedia, 22(1),* 148–159. 10.1109/TMM.2019.2922129.

Website of the Republic of Poland. Coronavirus: information and recommendations. www.gov.pl/web/koronawirus/dzialania-rzadu: accessed 06 August 2020.

Wiederhold, B.K. (2020). Using social media to our advantage: Alleviating anxiety during a pandemic. *Cyberpsychol. Behav. Soc. Netw., 23(4),* 197–198. 10.1089/cyber.2020.29180.bkw

World Health Organization (WHO). (2020a). WHO Director-General's statement on IHR Emergency Committee on Novel Coronavirus (2019-nCoV). www.who.int/dg/speeches/detail/who-director-general-statement-on-ihr-emergency-committee-on-novel-coronavirus-(2019-ncov): accessed 06 August 2020.

World Health Organization (WHO). (2020b). WHO Director-General's opening remarks at the media briefing on COVID-19 – 11 March 2020. www.who.int/dg/speeches/detail/who-director-general-s-opening-remarks-at-the-media-briefing-on-covid-19---11-march-2020: accessed 06 August 2020.

World Health Organization (WHO). (2020c). Novel Coronavirus (2019-nCoV) Situation Report-13. www.who.int/docs/default-source/coronaviruse/situation-reports/20200202-sitrep-13-ncov-v3.pdf: accessed 06 August 2020.

Worldometer. (2020). COVID-19 Coronavirus Pandemic. www.worldometers.info/coronavirus/: accessed 25 September 2020.

Yu, H., Xie, T., Paszczynski, S., Wilamowski, B.M. (2011). Advantages of radial basis function networks for dynamic system design. *IEEE Trans. Ind. Electron, 58,* 5438–5450. 10.1109/TIE.2011. 2164773.

6 Analysis of CCT through ANFIS for a Grid-Connected SPV System

Adithya Ballaji and Ritesh Dash

6.1 INTRODUCTION

A sun-dependent photovoltaic (SPV) grid network-related structure has a broad assortment of uses, starting from unconstrained system network-related applications. Extending the efficiency of sun-fueled cells furthermore fabricates the execution of SPV grid-integrated related systems. An increase in the SPV system-related grid network leads to an increase in quality issues to which it is related. Power quality issues include assortment in the voltage level at the place of point of common coupling, peak hour control dealing with cutoff and total harmonic distortion introduced in the electrical cable. As it is an intermittent wellspring of imperativeness, an SPV structure should be able to disengage itself in the midst of grid-integrated fault conditions. An islanding task is a basic component of a limitless grid-integrated associated system.

Various experts have focused on the current control loop and voltage control loop used as a fragment of the inverter regulator circuit to control assumptions about grid-connected PV system. Estimations take into account that direct and nonlinear frameworks are generally suitable for illustrating the current regulator. In any case, these regulators require a quick and focal point-based logical assessment for their

Email: Indiardasheee@gmail.com

DOI: 10.1201/9781003216278-6

execution and plan. Again, from a security point of view these regulators do not show elegant execution in the midst of transient and grid instability conditions, for instance, single line to ground fault condition. Clever methodologies, for instance, fuzzy and neural network-based regulators, may be associated with these issues. The advantages of an artificial intelligence (AI) system over standard techniques are that they do not need detailed mathematical examination for their implementation and accordingly are easy to realize, and SPVs are robust in the midst of faults in a grid-integrated network.

Artificial neural network techniques are being used because their controller design configuration enables their adaptability to the condition to which they are related. The development of methodology associated with this controller drives the controller to continue like a human neural control action where it applies its experience to find a decision under fluctuating grid conditions. Inverter current control methods were selected for the present circumstances as it interfaces both renewable power sources and that of the grid network for control.

6.2 FUZZY CONTROL TECHNIQUES

This section describes a fuzzy logic controller, which works under both varying and disproportionate conditions. This is possible in view of the fact that mathematical induction was not applied. A nonlinear framework can be best dealt with by a fuzzy reasoning regulator. Here the execution of the regulator is attempted under both changed and inconsistent conditions. Despite the current controlled fuzzy logic controller, this part moreover familiarizes another controller with control of the direct current (DC) capacitor voltage, over the inverter input terminal.

6.2.1 Proposed Control Strategies

The proposed control strategies for the current controller are shown in Figure 6.1.

The proposed current controller contains two main parts: (1) DC voltage to be maintained transversely over the capacitor; and (2) coupling inductor (Lf). DC voltage needing to be maintained at the input of the inverter is controlled by an outer voltage-controlled loop and a significant gate signal is delivered by internal current control loop. A coupling inductor is used to smooth the swell of consonant current injected from the inverter to the organization. In the proposed work, a fuzzy reasoning regulator-based pulse width modulation (PWM) controller is used to make the required switching signal for the inverter. The switching signal thus generated is upgraded and sent to the PWM converter for network synchronization. The DC interface voltage is thus kept predictable by a fuzzy reasoning controller.

6.3 DESIGN OF THE FUZZY LOGIC CONTROLLER FOR THE INNER CURRENT CONTROL LOOP

The block diagram of the fuzzy reasoning controller is shown in Figure 6.2.

The proposed fuzzy reasoning controller has two data sets: (1) current error; and (2) change in current error. With the specific objective of covering the entire extent of error, the proposed fuzzy reasoning controller has five membership functions (MFs).

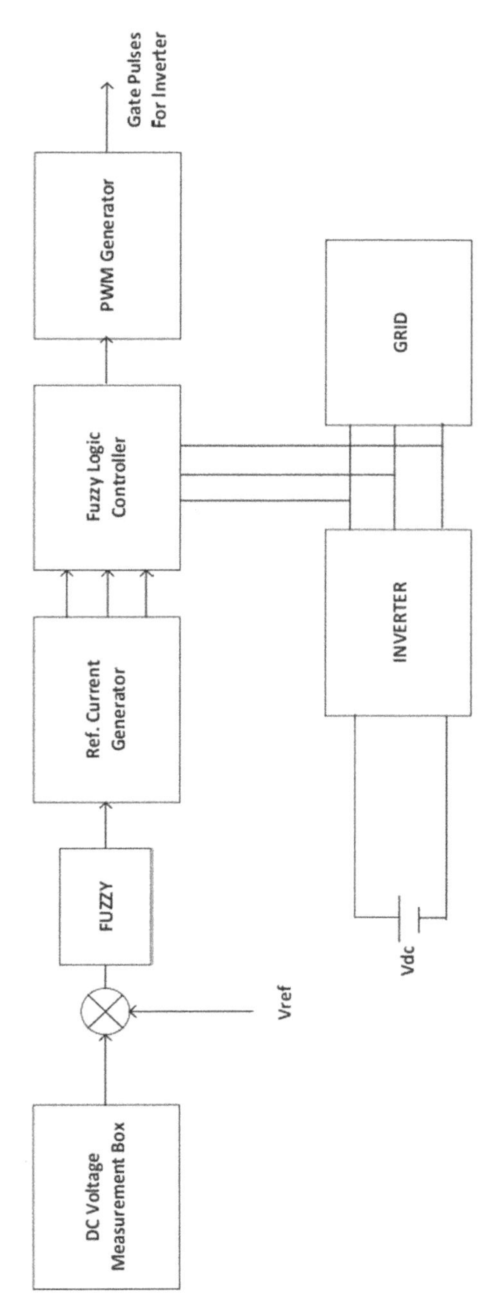

FIGURE 6.1 Schematic strategies of the current controller. DC, direct current; PWM, pulse width modulation.

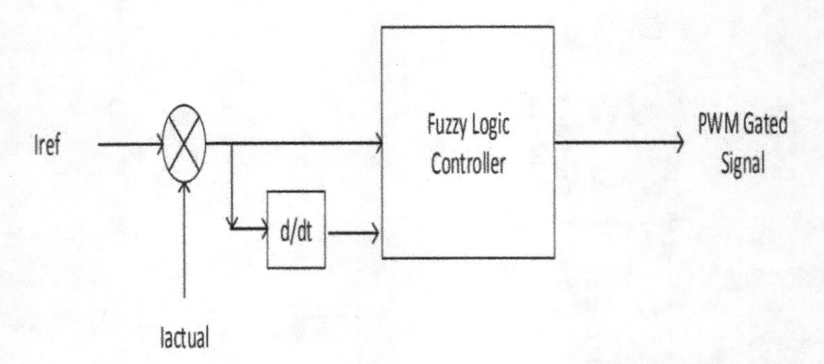

FIGURE 6.2 Schematic diagram for the fuzzy reasoning controller. PWM, pulse width modulation.

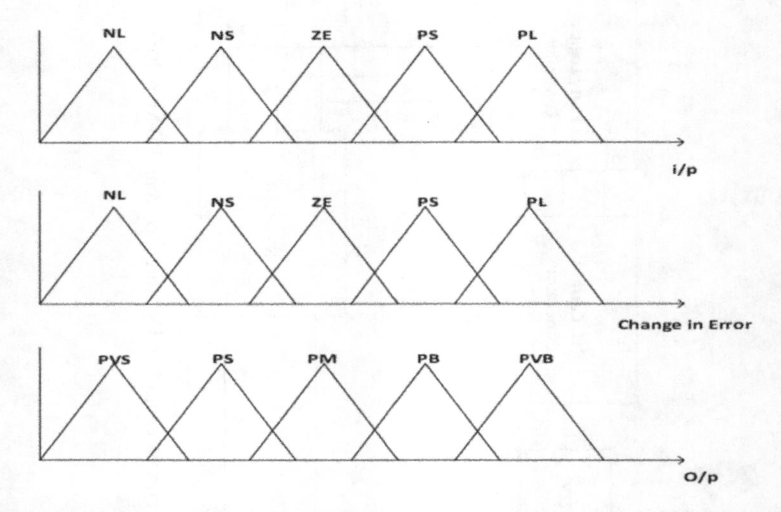

FIGURE 6.3 Membership function (MF) for input (i/p), change in error, and output (o/p) for current controller.

These MFs are normal for both input and output. Three-sided participation work cover the extent of the factors; with two input and five variable limits, 25 level sets can be achieved. Enrollment capacities for data and output factors are shown in Figure 6.3.

6.3.1 Design of the Fuzzy Logic Controller for the Outer Voltage Control Loop

DC voltage at the input of the inverter is constrained by a capacitor. This capacitor regularly gives consistent voltage and supplied the real power demand by the grid. Under continuing movement conditions additional real power demand should be identical to control supply by the inverter with some proportion of additional power

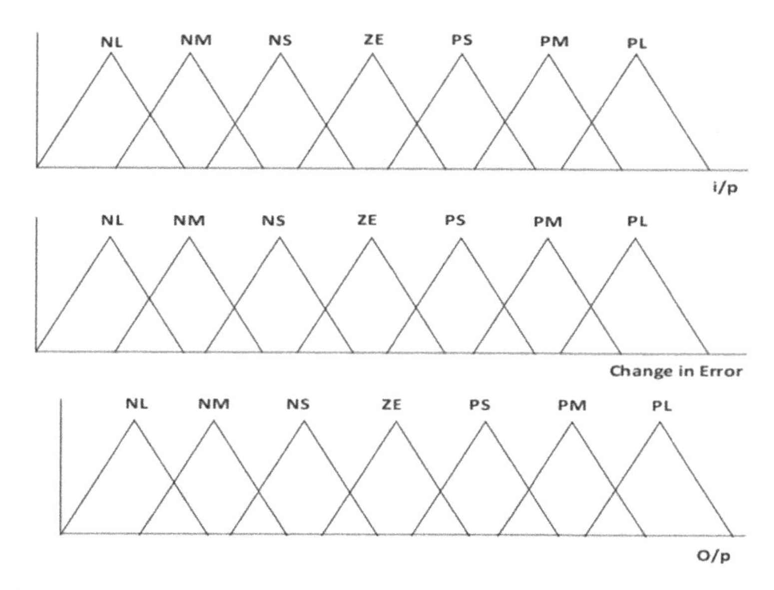

FIGURE 6.4 Fuzzy reasoning controller representation of etymological variables.

problems. A fuzzy reasoning controller is executed to keep up a consistent voltage over the capacitor at the input of inverter. Fuzzy reasoning control factors are selected in agreement with the dynamic execution of the controller. Variations in error are believed to be as a result of controller. Along these lines, real power demand present at the grid output side is believed to be the output of the fuzzy reasoning controller. Both input and output are addressed by seven semantic elements. These etymological factors are:

1. Negative large (NL)
2. Negative medium (NM)
3. Negative small (NS)
4. Zero (ZE)
5. Positive small (PS)
6. Positive medium (PM)
7. Positive large (PL)

Fuzzy reasoning controller representation of etymological variables is shown in Figure 6.4.

6.3.2 Design of the ANFIS Controller

Artificial neural fuzzy inference system (ANFIS) has a Takagi–Sugeno fuzzy induction structure. Output of ANFIS ordinarily uses a mixture of input factors for certain consistent variables. These linear joined components are generally viewed as weight

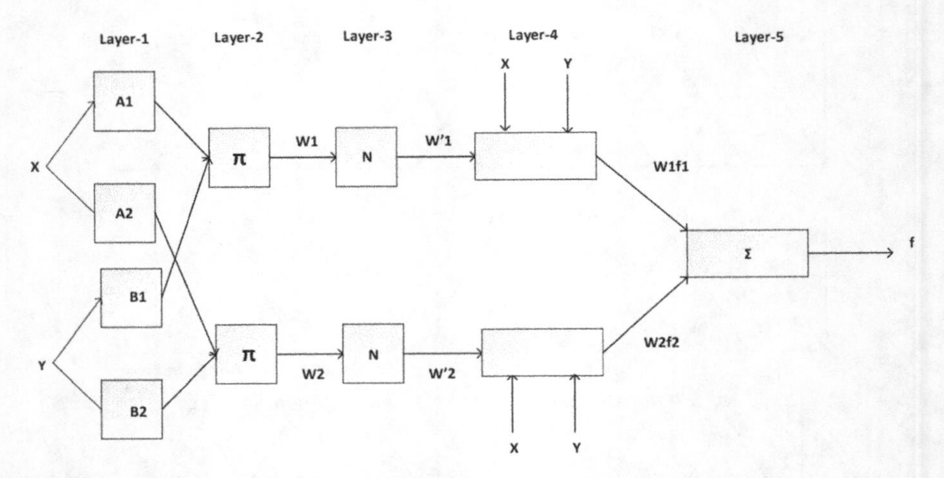

FIGURE 6.5 Basic artificial neural fuzzy inference system (ANFIS) structure.

work. The last output of the ANFIS structure is regularly a weighted total ordinary of the data changed into output. The central ANFIS structure involving two data points, for instance, x and y, and one output z is presented in Figure 6.5.

The Takagi–Sugeno fuzzy inference system rule for an if and then statement is as shown below:

Rule 1: If x is A1 and y is B1 then $f1 = P1x + q1x + r1$
Rule 2: If x is A2 and y is B2 then $f2 = P2x + q2x + r2$

Layer 1

For every node i, in this layer the square node function becomes:

$$O1,i = \mu A,i(x) \text{ where } i = 1,2$$
$$O1,i = \mu B,i(y) \text{ where } i = 1,2$$

Here x refers to commitment to junction i or layer 1 and "A" and "B" refers to etymological components associated with that particular junction. The phonetic variables can be either a three-sided or Gaussian proposed work. Boundaries attributed to layer 1 are named prelude boundaries.

Layer 2

The node level for layer 2 is π. The output of layer 2 is the product of all incoming signals.

$$O1,i = \mu A,i(x) \times \mu B,i(x) \text{ where } i = 1,2$$

The output of layer 2 represents the strength of fuzzy rule.

Layer 3

The node level for this layer is *N*. It represents the ratio of strength of fuzzy rule to sum of strength of fuzzy rules.

$$O3,i = wi = wi/(w1 + w2) \text{ for } i = 1,2$$

Output of layer 3 represents the normalized strength of fuzzy rule.

Layer 4

This layer is referred to as the adaptive node. The node function for layer 4 can be written as:

$$Oq,i = wifi = wi(p1x + qiy + ri)$$

Layer 5

The output of layer 5 represents the weighted sum average of the linearly variable input. The output equation for its layer can be written as:

$$Oc,i = \Sigma wifi$$

This confirms that ANFIS is similar to that of the Takagi–Sugeno-type fuzzy inference system.

6.4 ANALYSIS OF RESULTS

The fuzzy inference system file thus generated is exported to workspace for further analysis of the performance of the controller under different grid conditions such as balanced and unbalanced. Real and reactive power delivered by the ANFIS-based controller under balanced conditions is shown in Figures 6.6 and 6.7.

As shown in Figure 6.6, real power exchanged between the renewable source and grid has undergone a transient change at 1.579 seconds because of change in load demand. Similarly, the system is stabilized at about 2.093 seconds after a short transient condition. This shows that system performance is much better to other fuzzy-based proportional integral (PI) controllers. Similarly to real power, reactive power exchanged between the two sources is almost negative such that the system is consuming excessive reactive power present in the grid.

The reference voltage is maintained at 500 V and that of DC voltage is found to be 507.49V; this shows that the ANFIS-based controller has better control over the external voltage control loop as compared to linear and fuzzy controllers. A constant modulation index of 100% is maintained throughout the simulation of controller (Figure 6.8).

Direct and quadrature axis current performance, as shown in Figure 6.9, are found to be satisfactory under balanced conditions of operation. Similarly, the maximum power point tracker performance is very fast with step change of 0.038 seconds between two states (Figure 6.10).

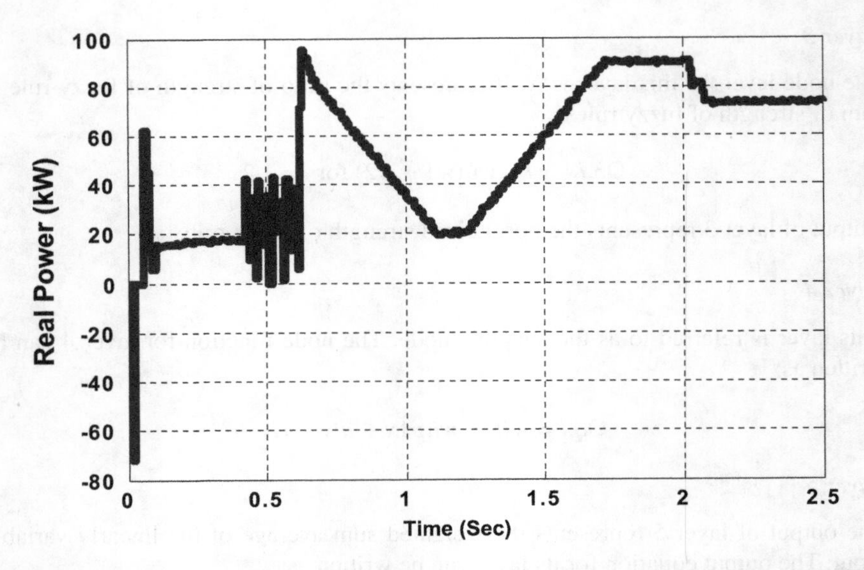

FIGURE 6.6 Real power delivered by the artificial neural fuzzy inference system (ANFIS) controller.

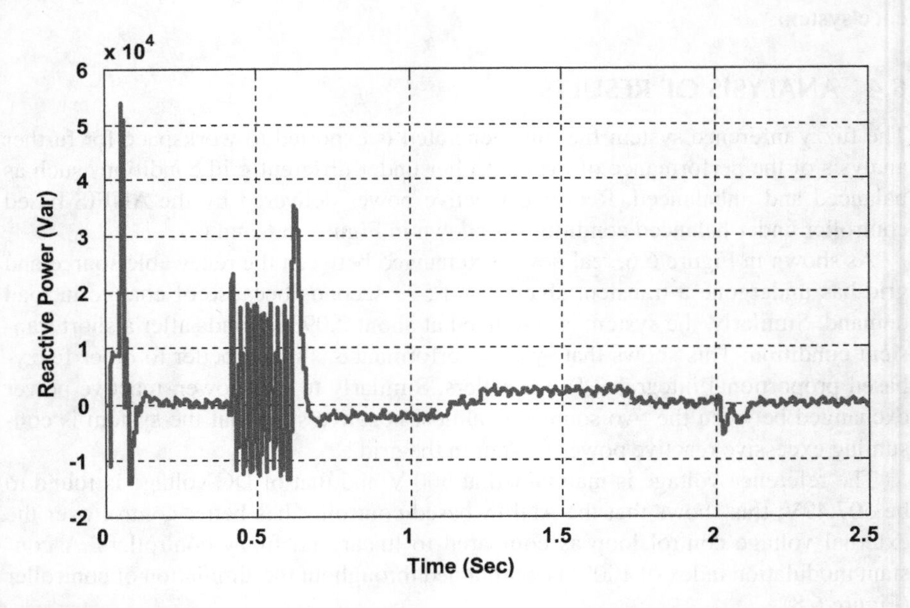

FIGURE 6.7 Reactive power exchanged between renewable source and grid.

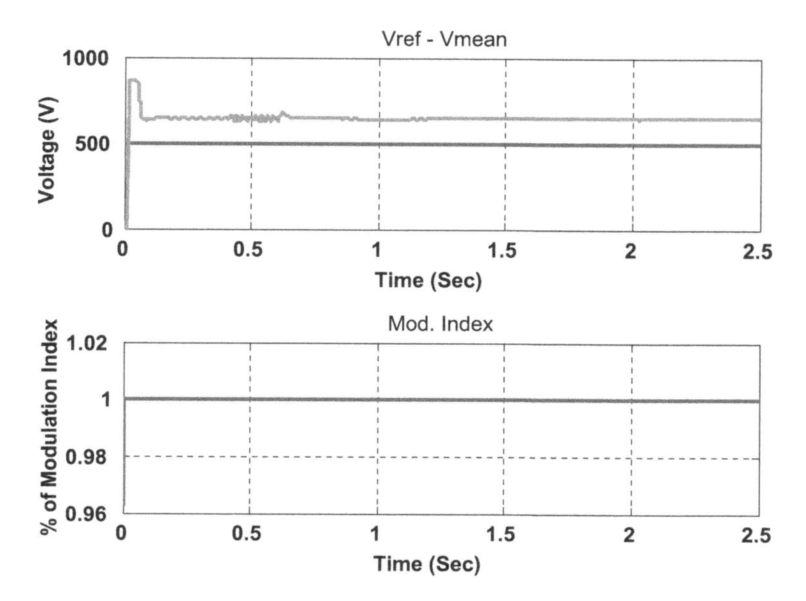

FIGURE 6.8 Reference voltage and modulation index.

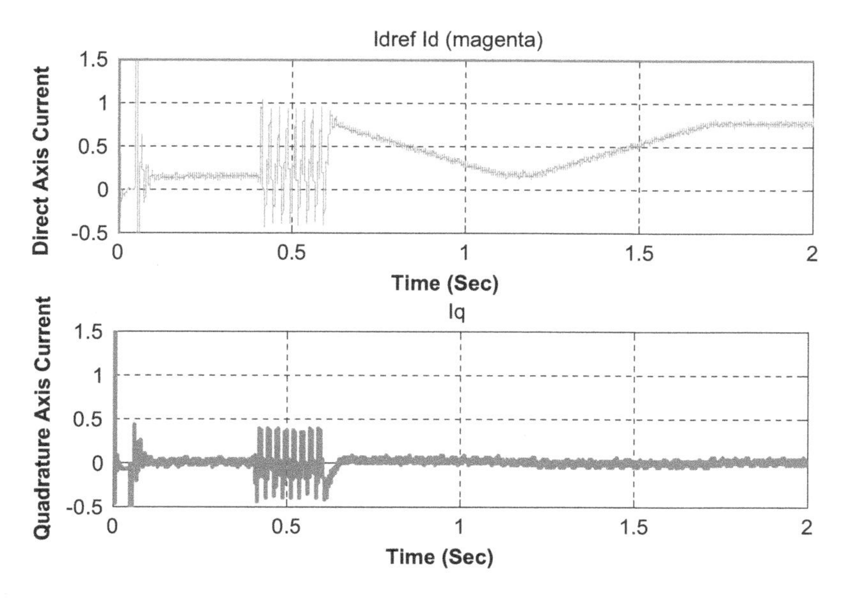

FIGURE 6.9 Direct axis and quadrature axis reference current.

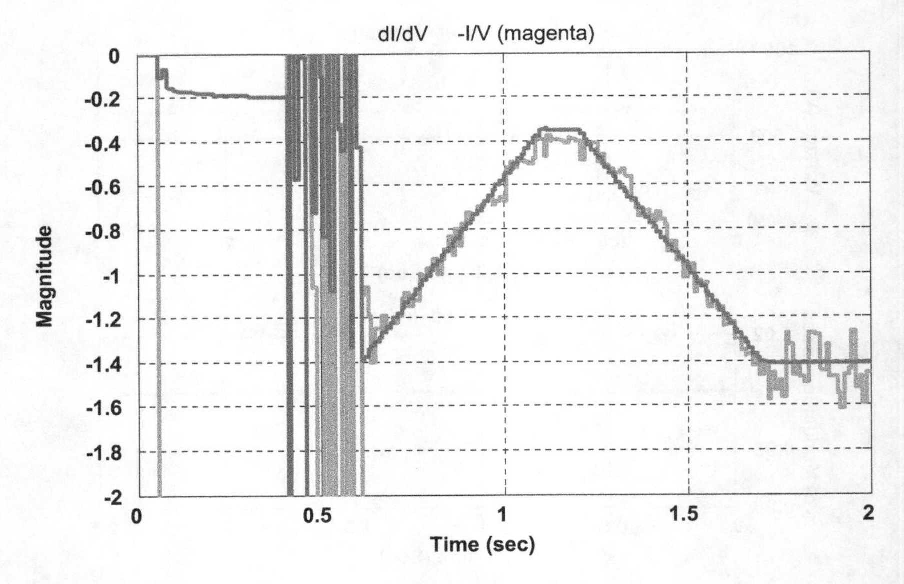

FIGURE 6.10 Maximum power point tracker (MPPT) performance.

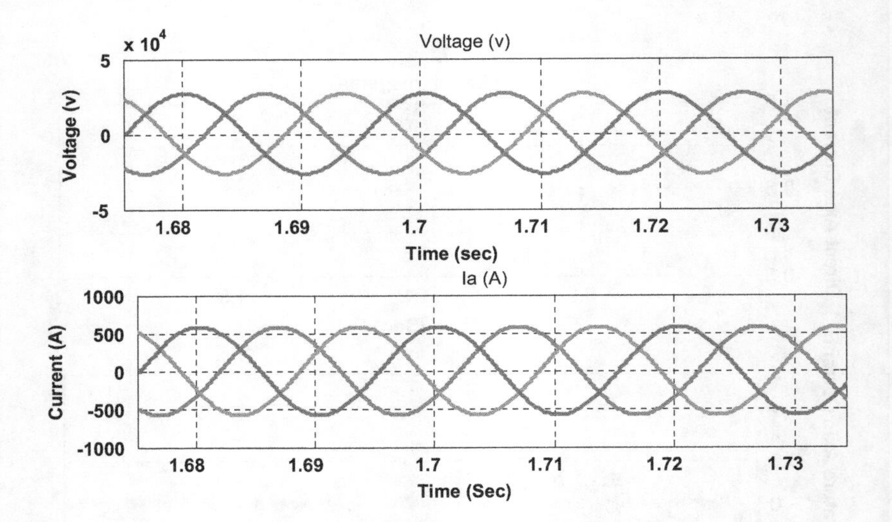

FIGURE 6.11 Voltage and current at point of common coupling (PCC).

Voltage and current status at the point of common coupling are shown in Figure 6.11. From this figure it was found that a voltage of 33 kV and current rating of 515 Å are delivered by the source into the grid.

6.5 CONCLUSION

A fuzzy logic controller provides better performance compared to a PI controller. This is because the fuzzy controller has greater flexibility by providing a larger area for data recognition and interpretation. However, this controller requires a lot of trial and error to achieve the best results. Unlike the fuzzy logic-based controller, ANFIS does not require trial and error to arrive at a particular result. From the model it can be found that mean squared error (MSE) for the Levenberg–Marquardt model is less, as obtained from sample 2. However, regression analysis shows that the Levenberg–Marquardt model for sample 4 has the least error. MATLAB modeling based on sampling percentage has been demonstrated for the current loop under balanced and unbalanced conditions. The result obtained for both balanced and unbalanced conditions performed better than linear controllers such as PI and fuzzy. ANFIS has better tracking capability over other algorithms.

BIBLIOGRAPHY

1. Milan Pradanovic, Timothy Green, "Control and filter design of three phase inverter for high power quality grid connection", IEEE Transactions on Power Electronics, vol. 18, pp. 1–8, January 2003.
2. Jyh-Shing Roger Jang, "ANFIS: adaptive-network-based fuzzy inference system", IEEE Transactions on Power Electronics, vol. 23, pp. 665–685, May/June 1993.
3. A. Kusagur, S.F. Kodad, B.V. Sankar Ram, "Modeling design and simulation of an adaptive neuro-fuzzy inference system (ANFIS) for speed control of induction motor", International Journal of Computer Applications, vol. 6, no. 12, pp. 29–44, September 2010.
4. G. Adamidis, G. Tsengenes, "Three phase grid connected photovoltaic system with active and reactive power control using instantaneous reactive power theory", Proceedings of the International Conference on Renewable Energies and Power Quality, pp. 8–16, March 2010.
5. Mateus F. Schonardie, Denizar C. Martins, "Application of the dq0 transformation in the three phase grid connected PV system with active and reactive power control", Proceedings of the Power Electronics Specialists Conference, pp. 1202–1208, June 2008.
6. Shuitao Yang, Qin Lei, Fang Z. Peng, Zhaoming Qian, "A robust control scheme for grid-connected voltage source inverters", IEEE Transactions on Industrial Electronics, vol. 58, pp. 1002–1009, January 2011.
7. Bo Yang, Wuhua Li, Yi Zhao, Xiangning He, "Design analysis of a grid-connected photovoltaic power system", IEEE Transactions on Power Electronics, vol. 25, pp. 992–1000, April 2010.
8. Jeyraj Selvaraj, Nasrudin A. Rahim, "Multilevel inverter for grid-connected PV system employing digital PI controller", IEEE Transactions on Industrial Electronics, vol. 56, pp. 149–158, January 2009.
9. Zhilei Yao, Lan Xiao, "Control of single-phase grid-connected inverters with non-linear loads", IEEE Transactions on Industrial Electronics, vol. 60, pp. 379–383, April 2013.

7 Frequency Analysis of Human Brain Response to Sudarshan Kriya Meditation

Rana Bishwamitra and Hima Bindu Maringanti

7.1 INTRODUCTION

The various brain states affecting the human mental condition, that in turn are externally perceived as behavior, are explained below.

7.1.1 Brain State/Human State of Mind

In an electroencephalogram (EEG), four types of brain waves exist, classified according to their frequency range: alpha (α) 8–13.9 Hz; beta (β) 14–30 Hz; theta (θ) 4–7.9 Hz; and delta (δ) 1–3.9 Hz. Irrespective of brain function, brain waves generate information: alpha waves show relaxation (SE-Con) and beta waves indicate cognitive skills (Con). Likewise delta waves reveal the learning ability of the brain, and theta waves give information on the self-management of memory (SU-Con) [1, 2].

Email: bishwamitra.rana@gmail.com; profhbnou2012@gmail.com

DOI: 10.1201/9781003216278-7

7.2 LEVELS OF CONSCIOUSNESS

Consciousness is the mental and awakened state of a human, which enables a person to become aware of their surroundings and accordingly action takes place. In order to know the mental and psychological effects required to analyze these states in each level of consciousness [2, 3], the levels of consciousness are described.

Semi-consciousness (SE-C) is the first level of consciousness, where awareness is relatively low and in this state dreams are made. Generally, semi-consciousness is represented as the receptacle of all past experiences, impressions left on the mind by those experiences or positive feedback from those impressions.

The *super-consciousness* (SU-C) level seem all things as a part of a whole. It is a conceptual level. At this level problems and solution are seen as one and the same. Regular practitioners or regular 'Sadhana' may feel bliss at this level, called 'Para- mananda'.

At the *unconscious* (UN-C) level, individual behavior changes as a result of the association of different experiences. Naturally, it occurs in sleep mode. Long-term memory is established at this level.

Generally, we receive guidance in the *consciousness* state, finding the best solution from many in the conscious state where experiences are derived from the unconscious mind [3]. The semi-conscious is in some ways close to the super-conscious level, where real intuition resides; further studies are required to demonstrate both these states.

7.3 DISCUSSION OF TECHNIQUES AVAILABLE TO RECORD THE HUMAN BRAIN RESPONSE

According to cognitive neuroscientists, the various brain imaging tools available to help understand brain activity include EEG, magnetic resonance imaging (MRI), functional MRI (fMRI), positron emission tomography (PET), and magnetoencephalography (MEG)h [4–7]. Specific tools is designed in specific techniques for image processing.

The EEG measures the electrical activity of the brain using electrodes. Electrodes are placed over the scalp to record activity in different regions of the brain, such as the frontal, parietal, occipital, and temporal lobes. In PET scans, a tracer substance is attached to radioactive isotopes and injected into the blood. When active, blood (which contains the tracer) is sent to deliver oxygen which helps to create visible spots that are picked up by detectors and used to create a video image of the brain while performing a specific task.

MRI uses hydrogen proton atomic nuclei within the cell body to generate a picture of the brain. Nuclei change their axis; in rotation the radiofrequency signal is observed until it reaches its former state. MRI can be used in utero and in infants and provides continuous imaging across the development spectrum.

fMRI measures the behavior of the brain while neural activity is changing. When an individual performs a task, the blood flow in a particular area enables identification of the location where the action is taking place.

MEG provides better spatial resolution, although it uses a radioactive tracer to record the cerebral magnetic field.

A recording with a super-conducting quantum interference device (SQUID) it can map brain activity by electromagnetic recording, producing electrical currents in the brain. As a result, images have better resolution. In contrast, photon migration tomography (PMT), otherwise known as near-infrared spectroscopy or optical imaging, is a new procedure for quantifying cortical activity. It assesses the scattering of near-infrared light from brain tissue. Transcranial magnetic stimulation (TMS) is a non-invasive and painless technique that excites neurons using time-varying magnetic fields.

7.4 OBJECTIVE

The objective of this work was to record EEG signals of the meditative brain(s) and analyze them to study the effect of meditation on brain activity.

7.5 LITERATURE SURVEY

The authors analyzed subjects' state of mind and observed the excitation of brain waves with Sudarshan Kriya (SKY) through EEG signals. Three phases of recording (pre-, mid-, and post-) were carried out and a common pattern was identified. Modified multifractal detrended fluctuation analysis (MFDFA) and multifractal detrended fluctuation analysis (MFDA) were applied as baseline analysis [8].

The authors established brain wave analysis pre- and post-musical session of EEG signal for validation and proposed a cognitive model for positive therapeutic effects on conscious state [1, 2].

The authors looked into the depth of varied image techniques applied on the brain in terms of development of multi-modal data integration. Analysis was based on risky decision making and illustrated how neuroimaging techniques can be applied to further research in advance [4].

The authors discussed multiple clinical problems related to the brain. EEG analysis was required to diagnose brain disease and the authors suggested that high-amplitude wave forms lead to seizure activity [7].

Some authors concluded that stress is a major factor disturbing the ongoing process of brain, and discussed factors involved in stress management along with how to reduce the tendency towards stress [9].

Some authors focused on mindfulness meditation in stress reduction. Structural changes on the EEG signal during meditation were discussed and different types of meditation such as Raj yoga and zen and SKYmeditation were claimed tol give positive psychological effects in experimental work [10].

The authors synthesized the EEG signal frequencies of SKY and its effects during meditation. A total of 20 subjects, of whom 11 were male and 9 female, were studied. The recorded signal was passed through a band pass filter at a frequency 0.5–40 Hz after normalized time series data, and the frequency analysis tool MATLAB was used [11].

The authors provided a comprehensive review of the effects of SKY by applying statistical and determination tests on the EEG signal. This results controlled stress regulation and improved cognitive level [12].

Here authors traced EEG brain waves during SKY meditation. Spectral analysis was carried out using quantitative analysis. For this purpose, the EEG data from 43 SKY practitioners were collected and compared with regard to stress and its relief pre- and post-session. It was concluded that long-term SKY practitioners had improved energy levels compared to others [13].

The authors identified the use of SKY breathing techniques by taking the pattern length of the subject and proposed a neurocognitive model, dividing different practices into their common elements for future research direction [14].

The authors analyzed the effects of SKY on the recorded EEG signal of 50 subjects. This was classified into two categories, practitioner and non-practitioner, based on standard deviation and decreased in control group. After a 3-month observation of pre- and post-kurthosis values, the control group was higher than the study group [15].

7.6 CONTRIBUTORY WORK

The present work aimed to trace brain waves and their response in terms of rhythmic breathing, founded by Sri Sri Ravi Shankar, Art of Living. The EEG signal was recorded subjects who were well trained in this technique and who had been in practice for the last 10–15 years. On that basis, experimentation has been considered for experimental work, to drive the baseline standard. The EEG signal was captured at periods of around 15 minutes and then a comparison was made of frequencies of alpha and theta amplitude and commonality patterns occurring in the brain identified. The common region excitation was interpreted and final inferences drawn. A mathematical tool wavelet transform (WT) was applied to time-scale signal analysis and EEG signal was decomposed. Time series data analysis was performed using fast Fourier transform (FFT) in MATLAB to obtain the results, denoted as $[X_1, X_2, \ldots X_N]$. The continuous frequency band was sliced, i.e. flow to f high into k bins. The bins are used for the frequency range of each particle. For these bins, we have band = {0,5,4,7,12,30,50}. Power spectral intensity (PSI) for the kth bin is calculated as given by:

$$PSI_{k'} = \sum_{i=\left|N(f_{k/f_s}\right|}^{\left|N(f_{k+1}/f_s\right|} \left|Xi\right|, k = 1,2,..,K-1$$

where f_s is the sampling rate, and N is the series length. The discrete wavelet transform analyzes the signal and decomposes it into coarse approximations to obtain detailed information. The decomposition of signal into different frequency bands is done by successive low- and high-pass filtering. In this way, the time series data of alpha and theta brain waves were obtained and MFDFA analysis was performed corresponding to each experimental condition.

7.7 EXPERIMENTAL SETUP

The EEG signal was recorded of three subjects in an afternoon session at around 2 p.m. in an air-conditioned and shielded measurement cabin. Subjects sat in a comfortable chair; an EEG recording cap with electrodes (Ag/Cl sintered ring electrodes)

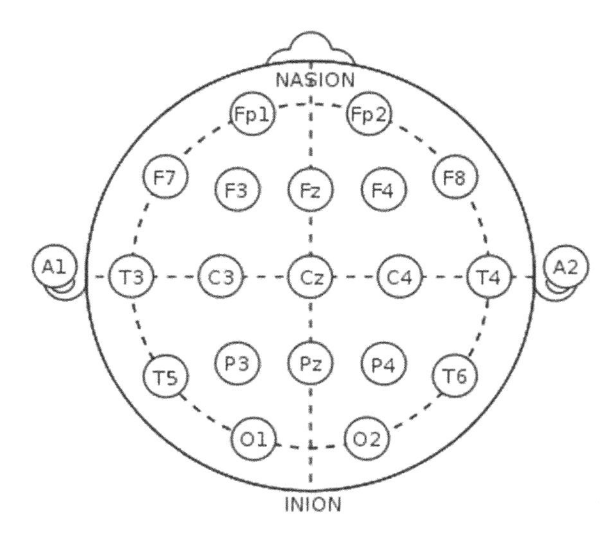

FIGURE 7.1 The position of electrodes according to the 10–20 international system.

was placed on the scalp. Impedances were checked below 5k Ohms. The EEG recording machine is operated at 256 samples with customized root mean square software (Figure 7.1). Set frequencies ranged from 0.5 to 35 Hz and raw EEG signals were synthesized as low- and high-pass filter. Electrical inference noise (50 Hz) was eliminated using notch filter; muscle artifacts were also eliminated through MG filter. A1 and A2 ear electrodes were linked together as reference. The same reference electrodes were used for all channels. After the electrodes were properly placed, each subject was recorded for 30 minutes and the following protocols were followed: the first 5 minutes were in a pre-meditative state; the second 15 minutes were in a meditative state; and the final 5 minutes were in a post-meditative state. The 15-minute period in the meditative state was divided into three time intervals, 2.15, 1.30, and 1.00 minute, to study the minute details of alpha and theta brain wave amplitude and their response. Three observations are shown as an example (Figure 7.2).

With respect to the analysis of the segment lasting 2.15 minutes, for subject NOU003, the dip in alpha asymmetry took longer to occur, whereas for subjects NOU001 and NOU002 it occurred much quicker. Similarly, theta asymmetry for subjects NOU001 and NOU002 occurred much quicker compared to subject NOU003. This indicates that the conversion of brain state to focused meditation took place at a much quicker rate for subjects NOU001 and NOU002 compared to subject NOU003.

In the case of the 1.30 minute observation period of subject NOU002, the dip in alpha asymmetry took longer to occur, while that for subjects NOU001 and NOU003 occurred at quicker. Similarly, the theta asymmetry index was higher in subjects NOU003 and NOU002 and occurred much quicker in comparison to subject NOU001. This indicates that the conversion of brain state to focused meditation took place at a much quicker rate for subject NOU002 compared to subjects NOU001 and NOU003.

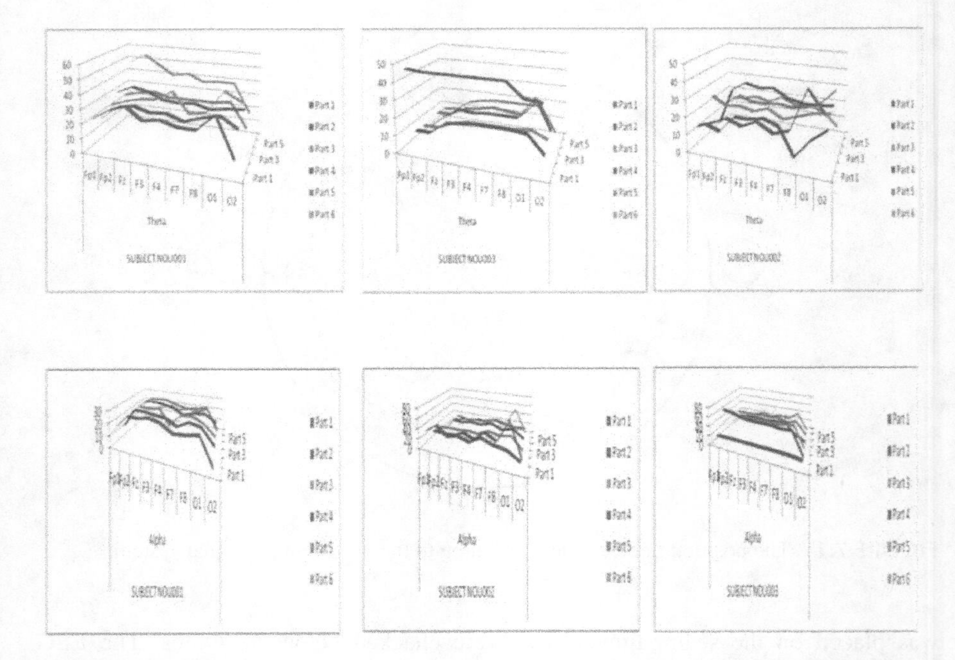

FIGURE 7.2 Analysis of alpha and theta power frequencies on subjects NOU001, NOU002, and NOU003 with time intervals of 2.15-, 1.30-, and 1-minute observations in a meditative state.

In the 1.00-minute period of observation, for subject NOU002, the dip in alpha asymmetry took longer to occur, while that for subjects NOU001 and NOU003 occurred at a much quicker response rate. Similarly, theta asymmetry index was higher in subjects NOU003 and NOU002 and occurred at a much quicker response rate in comparison to subject NOU001. This indicates that the conversion of brain state to focused meditation took place at a much quicker rate for subject NOU003 compared to subjects NOU001 and NOU002.

By subdividing the time studied into three different segments, commonality could be observed. In the frontal lobe, the asymmetry index for subjects NOU001 and NOU003 was on a much higher scale compared to the occipital lobe, for which the asymmetry index was on a much lower scale.

7.8 CONCLUSION

SKY is a novel approach that is undergoing extensive research to demonstrate it as an evidence-based therapy. SKY is an effective meditation, which is relied on not only for treating stress and anxiety, but also for post-traumatic stress disorder, depression, stress-related clinical illness, and substance abuse, and for rehabilitation of criminal offenders. It relieves stress and develops an individual's mind–body–spirit so that they can be happier, healthier, and possibly even long-lived. In the competitive modern world, in which stress and anxiety are part of everyday life, SKY-based therapy is very important. For all three subjects, a common trend was that, as the meditation period increased, the asymmetry index rose, signifying that the brain is going towards avoidance. For subject NOU002, the dip in alpha asymmetry took longer to occur, while that for subjects NOU001 and NOU003 occurred at a much quicker rate. This indicates that the conversion of brain state to focused meditation took place at a much quicker rate for subjects NOU001 and NOU003 compared to subject NOU002. In the frontal lobe, the asymmetry index was at a much higher scale as compared to the occipital lobe, which has the asymmetry index on a much lower scale. Hence, we find that the baseline could be drawn from subjects 001 and 003, with subject 002 responding slightly differently in comparison. However, the duration of deep meditative state in all three cases was around 15 minutes, a common pattern inferred from this experiment. In both frontal and occipital portions, suppressed alpha power activity is noticed in the right hemisphere which denotes focused attention and visual memory tasks during the period of meditation in the subject, indicating that the subject is experiencing visualization, a usual phenomenon amongst meditators. While in the frontal lobe, the right hemisphere has consistently low alpha power, the right occipital lobe has consistently varying periods of alpha power, indicating variations in the visual domain of the participant. In the pre-frontal region, however, we see that alpha power suppression is more prevalent in the left hemisphere, indicating positive valence during the meditational stages. The alpha power was found to be significantly higher in the frontal region as compared to the occipital regions, indicating and thus confirming the occipital region is much calmer.

7.9 FUTURE WORK

The above experiment should be repeated with naive users, people who have not done meditation at all, and then on new practitioners of SKY. These studies could be compared to the baseline signals inferred above for the level of achievement of that mindful state; the means of achieving the calibrated standard (as in baseline) would be suggested to be regular and repeated practice of SKY as the solution. In addition, the size of the expert long-term practitioner group could be increased to get the best and standard calibration. Moreover, the subjects themselves could be utilized to validate the inferences after signal analysis.

ACKNOWLEDGMENTS

Our sincere thanks are due to Prof. Deepak Gupta, Dr. Ranjan Sengupta, and Dr. Shankha Sanyal for extending all types of cooperation to conduct the experimental work at the C.V. Raman Centre for Physics and Music, Jadavpur University.

REFERENCES

1. Hima Bindu M, Consciousness and Music, Proceedings of 22nd International Symposium on Frontiers of Research in Speech and Music (FRSM 2015), pp. 123–127, November 23–24, 2015, Indian Institute of Technology (IIT), Kharagpur.

2. Rana B, Hima Bindu M, Music as a Means of Consciousness Awakening, 22nd International Symposium on Frontiers of Research on Speech and Music (FRSM2016), 11–12 November, North Orissa University.
3. Rana B, Hima Bindu M, Simulation of Behavior of Neurotransmitters, National Conference on Current Trends in Computing (NCCTC 2014), pp. 67–72, 23–24 March, 2014.
4. Gui X, Chuansheng C, Zhong L, Qi D, Brain Imaging Techniques and Their Applications in Decision Making Research, Acta Psychologica Sinica, 2010, 42(1): 120–137, doi:10.3724/SP.J.1041.2010.00120
5. Brain Imaging Techniques. Available online at: https://courses.lumenlearning.com/boundless-psychology/chapter/brain-imaging-techniques: accessed 14 September 2021.
6. Studying the Human Brain. www.news-medical.net/health/Studying-the-Human-Brain.aspx: accessed 14 September 2021.
7. Alan W, John W, Electroencephaloghaphy, in: J. Kreutzer et al. (eds.), Encyclopedia of Clinical Neuropsychology, Springer International Publishing, 2016, DOI 10.1007/978-3-319- 56782-2_24-3
8. Rana B, Hima Bindu M, Study of Long Term Effect (LTE) of Sudarshan Kriya (SK) and Music Listening (ML) Using EEG Analysis of Human Brain Response (HBR), in 13th Western Pacific Acoustic Conference (WESPAC 2018), New Delhi, November 11–15.
9. Chaudhuri A, Impact of Stress and Stress Management Programmes on Health in Recent Years: A Review, International Journal of Research & Review, 2019, Vol. 6, Issue 12, December.
10. Seema S K, Sonali B K, Review on Effect of Mindfulness Meditation on Brain through EEG, International Journal of Engineering Research and General Science, 2016, Volume 4, Issue 4, July–August.
11. Yugandhara M, Prajakta F, Review Paper on Electroencephalographic Evaluation of Sudarshan Kriya, International Journal of Science and Research (IJSR), 2012, 2319–7064.
12. Sushil C, Jaiswal A, Singh R, Jha D, Mittal A, Mental Stress: Neurophysiology and its Regulation by Sudarshan Kriya Yoga, International Journal of Yoga, 2017, 10(2), 67–72, doi: 10.4103/0973-6131.205508
13. Chandana V, Kochupillai V, Quantitative Analysis of EEG Signal Before and After Sudharshana Kriya Yoga, International Journal Of Public Mental Health And Neurosciences, 2015, 2(2), August.
14. Richard PB, Patricia LG, Sudarshan Kriya Yogic Breathing in the Treatment of Stress, Anxiety, and Depression: Part I – Neurophysiologic Model, Journal of Alternative and Complementary Medicine, 2005, 11, 189–201.
15. Sharma H, Raj R, Juneja M, EEG Signal Based Classification Before and After Combined Yoga and Sudarshan Kriya, Neuroscience Letters, 2019, 707, 134300, doi.org/10.1016/j.neulet.2019.134300

8 Coordinated Control Action between a DFIG-Grid Interconnected System Using a PI-Based SVM Controller

Adithya Ballaji and Ritesh Dash

8.1 INTRODUCTION

The concept of the smart grid requires the interconnection of renewable sources to the conventional power system to provide all-round protection in terms of power security, power quality and reliability. Therefore it is required to operate a large number of wind farms, with a restriction on reactive power absorption. As a result, the reactive power demand by the wind generators must be filled by a local power electronic device that will coordinate with the wind farm as required. To address this issue the capability curve or p q curve is a primary requirement. In this chapter, the main objective is to achieve coordinated control between a wind farm and static synchronous compensator (STATCOM) using a support vector machine (SVM) algorithm and power capability curve.

Most wind farms use induction generators such as fixed-speed and variable-speed induction generators. However, advancement in the development of the generator has made it possible to run wind farms with doubly fed induction generators. The induction generator usually operates at a lagging power factor which is the main cause of the requirement of a large amount of reactive power in the power system network. The amount of reactive power required by an induction generator depends on the amount of real power is being delivered, which again depends on many factors. With a fixed-speed induction generator a capacitor bank is sufficient to deliver the requirement

DOI: 10.1201/9781003216278-8

of reactive power by induction generator. However, with a variable-speed induction generator, the demand for reactive power is not static but dynamic; that is, it depends upon the situation and the amount of real power to be delivered to the system. Most distribution systems have observed that the requirement for reactive power increases linearly with an increase in real power or in a stepwise manner in proportion to a change in real power demand. Therefore, any demand in reactive power can be fulfilled by interconnecting suitable power electronics-based devices at the point of common coupling or optimized location in coordination with the wind farm.

From a reliability and safety point of view, the interconnected wind farm must obey the grid code without disconnecting the wind generator from the grid irrespective of the real power generated by it. Different countries through their electricity boards have confirmed that wind generators must be connected to the power system for a sufficient time during the fault condition without hampering the frequency, reactive power, real power, and voltage at the point of common coupling. It is also relaxed for the wind generator that the wind generator control action may not be able to provide sufficient compensation against reactive power and real power as other base stations provide during a fault condition.

During the coordinator control action, the different devices connected to the power system need to be taken care of because this system can affect the operation of the protective relay and thereby the performance of the system during fault conditions, bringing into question reliability. Therefore, the feasibility of the present power transmission system should be evaluated, to ascertain whether it is sufficient to be equipped with devices in a shunt-connected manner. Several research papers have been published on this particular topic; it has been found that the interconnection of devices at the midpoint can affect the performance of the transmission line as well as relays. These devices usually affect the distance protection relay in terms of the measurement of phase and operating time, ultimately leading to underreach and overreach problems in distance protection.

An SVM is a computer algorithm that requires both statistical learning and adaptive computation. It is a supervised learning algorithm used to solve and classify the data in a noisy and complex medium involving pattern recognition. It is used to learn from data by analyzing the relationships between them. It is very efficient while dealing with data in higher dimensions and uses strong mathematical structures. Most of the optimization problem involves solving a very large complex problem and requires a large memory. Therefore, these problems need to be solved as simply as possible. A standard SVM uses a binary classifier to classify the data using a hyper-plane and thus separates the input and non-members in the data space. It maps the data into a higher dimension and separates them using a maximum-margin hyper-plane. The system automatically identifies the support vectors, i.e., a subset of training data to separate the linear combination of points. Based on these support vectors it decides to solve the convex optimization problem. In practice, most SVMs are used to solve non-linear problem; therefore in this chapter a non-linear problem has been addressed with linear SVM.

Solving linear equations using SVM requires three parameters: problem formulation, solver, and optimization. Here optimization increases the efficiency of the solver and overall performance of the system. Most of the SVM problem is formulated as a convex problem because all the optimized points can be regarded as the global minimum and the algorithm does not stick at any point. This can be achieved using two types of function: as loss function and regularization. Problem formulation for a particular case can be formed in two ways, i.e., either primal mode or dual mode. If the number of features is less, then primal is preferred over dual. Distributed and parallel computing are features used to solve big data analysis problem. Big data problems can be solved using a stochastic algorithm. Of the stochastic algorithms, mini-batch approach and coordinate descent algorithms were generally preferred because the first algorithm always removes the dependent variable and the second algorithm transforms a big problem into a small set of vectors where each feature is treated as a variable. Apart from the above algorithms, another algorithm called proximal algorithm was also used to solve complex problems which are a combination of mini-batch and coordinate descent algorithm.

8.2 PROBLEM FORMULATION AND SOLUTION

Problem formulation means designing the mathematical equation for the case under investigation. In SVM the problem consists of some mathematical objective function and constraints. As observed, linear SVM operates in input space and non-linear SVM operates in feature space. However, non-linear SVM can be solved by applying the Kernel function to linear SVM, as per Equation (8.1):

$$x_1^T x_2 \rightarrow K\left(x_1 x_2\right) \tag{8.1}$$

where x_1, x_2 denote input data point and K represents the Kernel function.

Let us consider there are "n" training data sets collected during a fault condition, i.e., the line to ground fault condition within zone 1 presented in a matrix format "A" with text and labeled by +1 or –1. A diagonal matrix θ consists of all elements with all diagonal elements as +1 or –1. Linear Kernel over matrix A has been applied to transfer the data from non-linear points to linear points. As mentioned earlier, the accuracy of the algorithm is best described by the robustness of the hypothesis. Here, it has been assumed that the hypothesis is a sigmoid type of function bounded by $0 \leq h_\theta(x) \leq 1$. From linear regression:

$$h_\theta\left(x\right) = \theta^T x \tag{8.2}$$

or:

$$h_\theta\left(x\right) = g(\theta^T x) \tag{8.3}$$

where g represents the sigmoid function.

Or:

$$h_\theta\left(x\right) = \frac{1}{1 + e^{-\theta^T x}} \tag{8.4}$$

So the cost function for the proposed SVM becomes:

$$J(\theta) = \begin{cases} -logh_\theta(x), & if\ y = 1 \\ -log(1 - h_\theta(x)), & if\ y = 0 \end{cases} \tag{8.5}$$

or:

$$J(\theta) = -ylog(h_\theta(x)) - (1 - y)log(1 - h_\theta(x)) \tag{8.6}$$

So for "*n*" number of data sets, the cost function becomes:

$$J(\theta) = -\frac{1}{m}\left[\sum_{i-1}^{n} y^{(i)} \log\left(h_\theta x^{(i)}\right) + (1 - y^{(i)} \log\left(1 - h_\theta x^{(i)}\right)\right] \tag{8.7}$$

Here, $h_\theta(x)$ represents the predicted value and can be described to coordinate the control action.

Proper classification is now required to apply regression analysis to predict the value of *Kp* and *Ki* in the proportional integral (PI) controller of the STATCOM to achieve coordinated control action; therefore, the objective function for the regression analysis is shown in Equation (8.8):

$$\min_{(\theta, y) \in R^{n+1+l}} y + \frac{1}{2}\|w\|^2 \tag{8.8}$$

s.t. $\theta(Aw - y) + y \geq 1$

This can be solved by applying the Lagrangian multiplier.

So applying the Lagrangian multiplier and its dual form, the equation can be further modified as:

$$\min_{0 \leq u \leq R^l} \frac{1}{2} u^T Q u - I^T u \tag{8.9}$$

where *u* represents the duality of the variable.

8.3 ANALYSIS OF RESULTS

The training status is shown in the Table 8.1.

Table 8.1 shows the test results for different percentages of training data set. Root means square error remains the same for 60%, 65%, 70%, 75%, 80%, and 85% of the training data. In this model, the best *R*-squared value is 0.94 for 60% of the training data (Figure 8.1).

Again the mean absolute error is about 1% for test sample 60% data and 85% data. And also the time taken by model no. 6 for the training of the data is 26.066 seconds as compared to 19.977 seconds for model no. 1. So the overall evaluation of the model

TABLE 8.1
Error Evaluation for 60–80% of the Training Data Set

(a) 60% of the training data set		*(b) 65% of the training data set*	
RMSE	0.03	RMSE	0.01
R-squared	0.94	R-squared	0.91
MSE	0.00	MSE	0.02
MAE	0.01	MAE	0.02
Prediction speed	–600 000 obs/s	Prediction speed	–410 000 obs/s
Training time	19.977 s	Training time	24.501 s
(c) 70% of the training data set		*(d) 75% of the training data set*	
RMSE	0.05	RMSE	0.03
R-squared	0.80	R-squared	0.89
MSE	0.00	MSE	0.00
MAE	0.02	MAE	0.00
Prediction speed	–630 000 obs/s	Prediction speed	–380 000 obs/s
Training time	21.929 s	Training time	21.0166 s
(e) 80% of the training data set		*(f) 85% of the training data set*	
RMSE	0.05	RMSE	0.01
R-squared	0.77	R-squared	0.93
MSE	0.00	MSE	0.00
MAE	0.02	MAE	0.01
Prediction speed	–760 000 obs/s	Prediction speed	–540 000 obs/s
Training time	1.9795 s	Training time	26.066 s3

RMSE, root mean squared error; MSE, mean squared error; MAE, mean absolute error.

is that model number 1, 6, or 2 is best fitted for the training data. In the present model, all three models have been evaluated to find the accuracy of the model (Figure 8.2).

Figures 8.3 and 8.4 show the performance of features for model 1 and 6. For both models the accuracy of prediction remains the same, i.e., 96.01%. However, model no. 1 fails to predict the gain parameter during severe grid disturbance; this is because the features selected for model no. 1 are not dynamic to the grid condition. Therefore, model no. 6 is found to be most suitable (Figures 8.5–8.7).

8.4 CONCLUSION

In this chapter the SVM-based algorithm has been discussed, as well as its application to the design of the parameters for PI controller used in the proposed coordinated control action of STATCOM and wind farm. The design of the PI controller has been used in many applications. The proposed method of tuning of the controller parameters is purely online and the system may pose an error in large variations in system signal. In the proposed control strategy, it was found that even if there is discrimination in the data, the control action is guaranteed and becomes robust in all

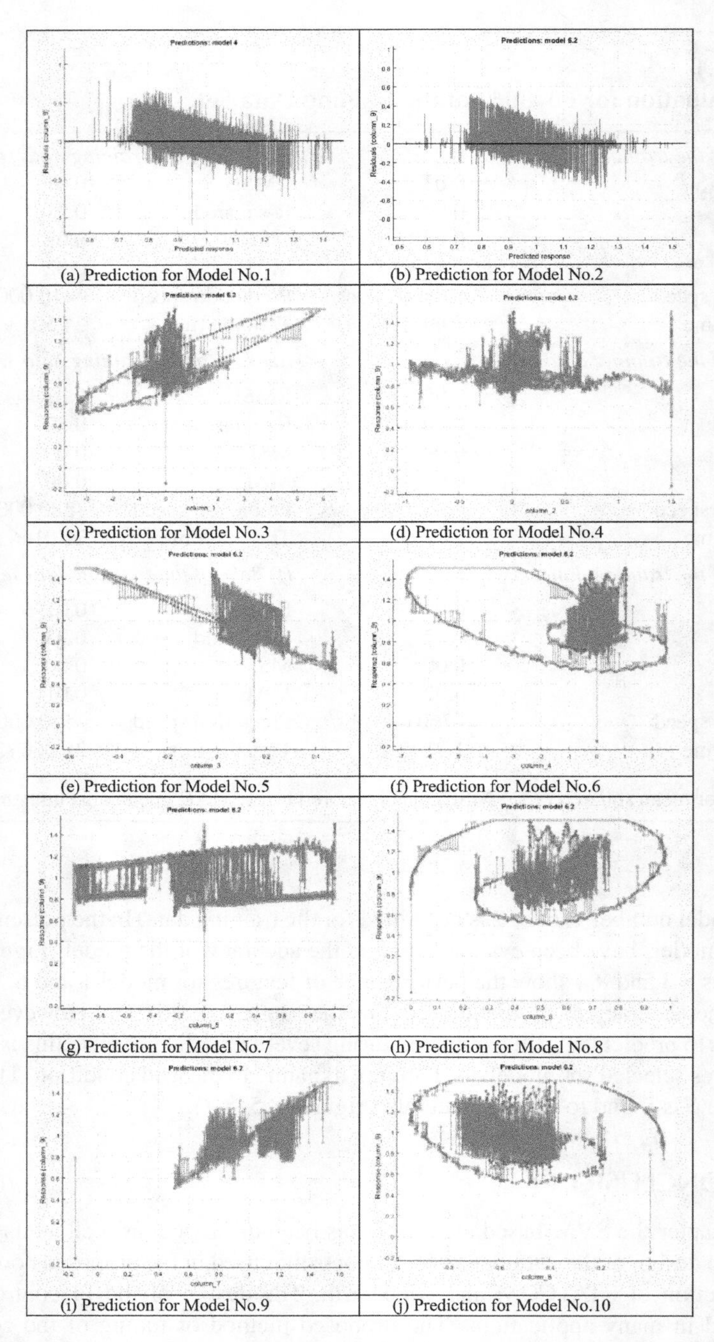

FIGURE 8.1 Prediction of (a) 60% of the training data set; (b) 65% of the training data set; (c) 70% of the training data set; (d) 75% of the training data set table; (e) 80% of the training data set table; and (f) 85% of the training data set.

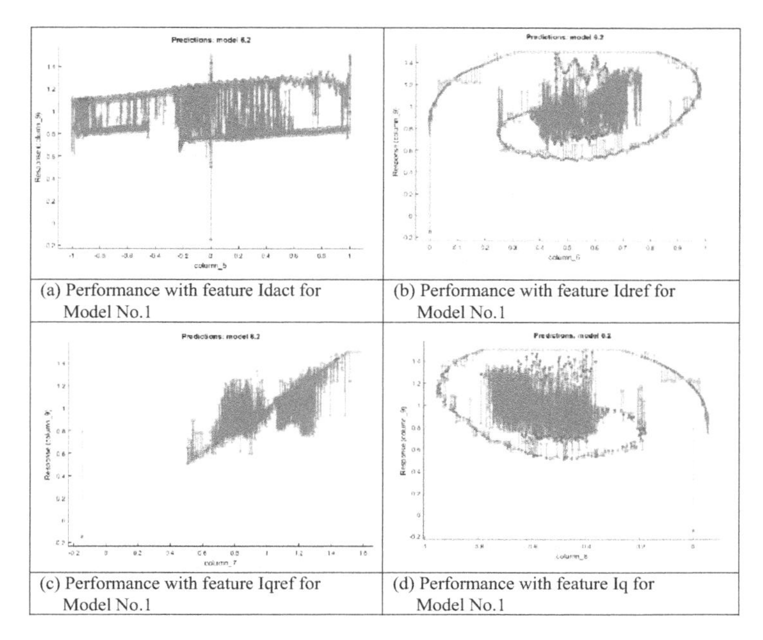

FIGURE 8.2 (a) Performance with feature Idact for model no. 1; (b) performance with feature Idref for model no. 1; (c) performance with feature Iqref for model no. 1; (d) performance with feature Iq for model no. 1.

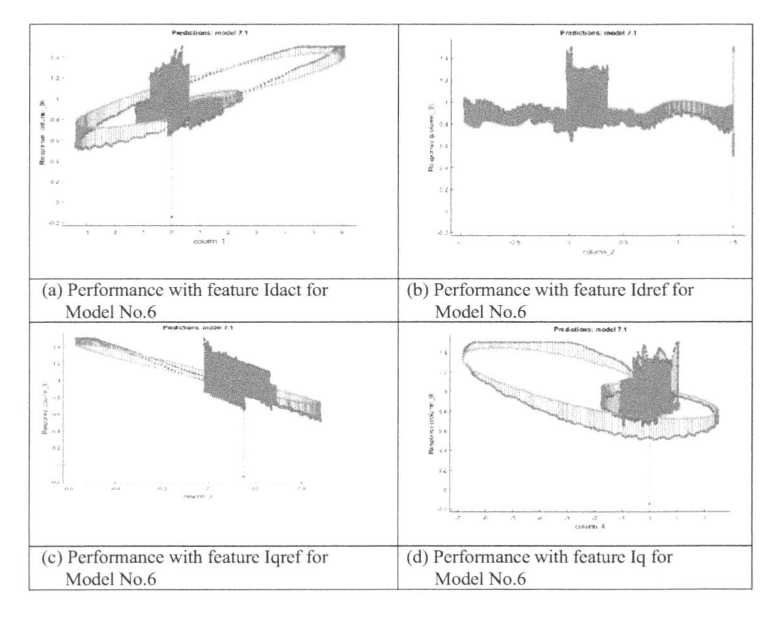

FIGURE 8.3 (a) Performance with feature Idact for model no. 6; (b) performance with feature Idref for model no. 6; (c) performance with feature Iqref for model no. 6; (d) performance with feature Iq for model no. 6.

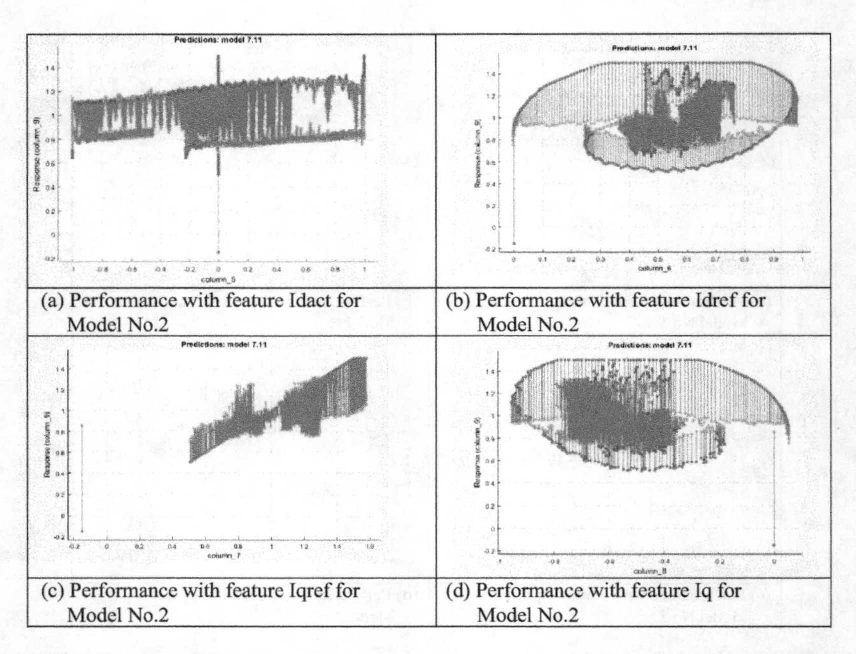

(a) Performance with feature Idact for Model No.2

(b) Performance with feature Idref for Model No.2

(c) Performance with feature Iqref for Model No.2

(d) Performance with feature Iq for Model No.2

FIGURE 8.4 (a) Performance with feature Idact for model no. 2; (b) performance with feature Idref for model no. 2; (c) performance with feature Iqref for model no. 2; (d) performance with feature Iq for model no. 2.

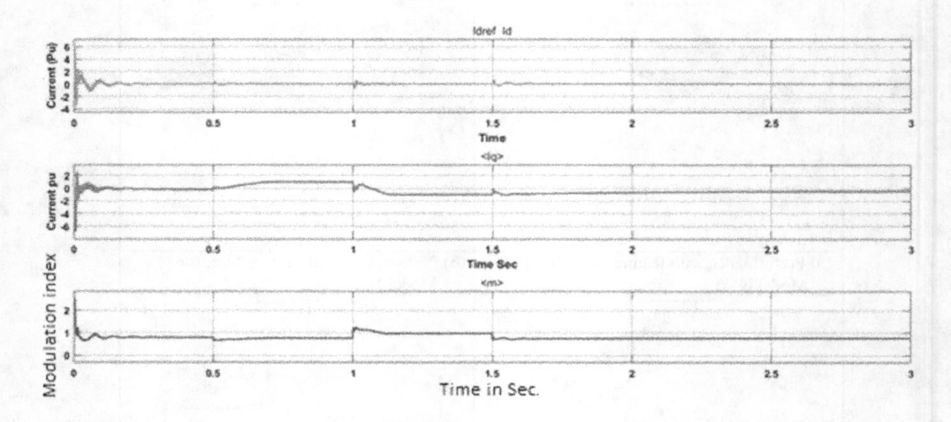

FIGURE 8.5 Id, Iq, and modulation index of the STATCOM control action as seen at proportional integral (PI) controller during line-ground (LG) fault.

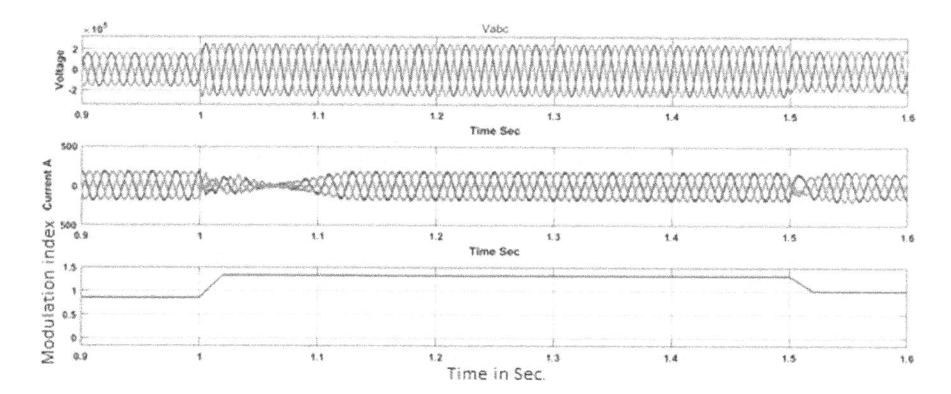

FIGURE 8.6 Point of common coupling (PCC) voltage and STATCOM terminal voltage during line-line-line-ground (LLLG) fault.

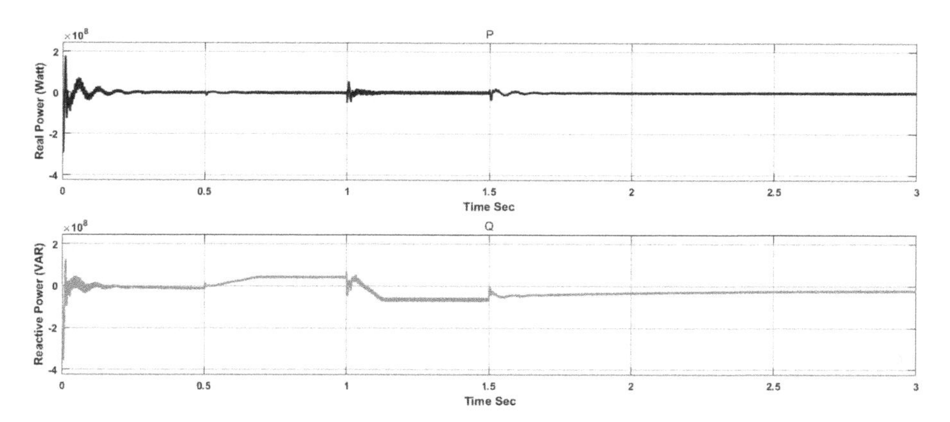

FIGURE 8.7 Real and reactive power exchanged between STATCOM and doubly fed induction generator (DFIG) during line-line-line-ground (LLLG) fault condition.

conditions because the algorithm has been prepared taken into consideration all the worst conditions. As mentioned, 60,000 training data points have been utilized with ten column features. Therefore, uncertainty in forecasting the gain margin for the PI controller was avoided.

ACKNOWLEDGMENT

The authors would like to thank the School of Electrical and Electronics Engineering, REVA, University, Bangalore for providing the necessary laboratory facilities and services for successful completion of this chapter.

BIBLIOGRAPHY

1. International Energy Agency (2017) Introduction and scope, in World Energy Outlook. IEA, Paris, pp. 32–62.
2. Global Wind Statistics 2017 (2018) http://gwec.net. Accessed 15 Jan 2018.
3. Installed Capacity in India 2017 (2019) www.cea.nic.in. Accessed 15 Jan 2019.
4. Ministry of New and Renewable Energy (2019) http://mnre.gov. Accessed 15 Jan 2019.
5. Nguyen HM, Naidu DS (2011) Advanced control strategies for wind energy systems: an overview. In: 2011 IEEE PES Power Systems Conference and Exposition, Phoenix, Arizona, pp. 1–8.
6. Baloch MH, Wang J, Kaloi GS (2017) A review of the state of the art control techniques for wind energy conversion system. Int J Renew Energy Res 6(4):1276–1295.
7. Vazquez Hernandez C, Telsnig T, Villalba Pradas A (2017) JRC Wind Energy Status Report – 2016 Edition. EUR 28530 EN. Publications Office of the European Union, Luxembourg. https://doi.org/10.2760/33253 5
8. Chang L (2002) Wind energy conversion systems. IEEE Canadian Rev 12–16.
9. Shanker T, Singh R (2012) Wind energy conversion system: a review. In: 2012 Students' Conference on Engineering and Systems, Allahabad, Uttar Pradesh, pp. 1–6. https://doi.org/10.1109/ SCES.2012.6199039
10. Vasipalli V (2015) Power quality improvement in DFIG system with matrix converter in wind energy generation with space vector control techniques. In: 2015 IEEE International Conference on Technological Advancements in Power and Energy, Kollam, pp. 73–78.
11. Khajeh A, Ghazi R, Abardeh MH, Sadegh MO (2018) Evaluation of synchronization and MPPT algorithms in a DFIG wind turbine controlled by an indirect matrix converter. Int J Power Electron Drive Syst 9(2):784–794.
12. Goudarzi WDZN (2013) A review on the development of wind turbine generators across the world. Int J Dyn Control 1(1):192–202.
13. Grid Code (2019) High and extra high voltage. www.eonnetz.com. Accessed 20 Feb 2019.
14. Khuntia GP, Jena R, Swain SC, Dash R (2020) FC-TCR based controller for SPV grid interconnected system, Materials Today: Proc 2020, ISSN 2214-7853.

9 Acceptance of New Schools in Semi-Urban Areas of India

An Application of Data Mining

Apurva Vashist and Suchismita Mishra

9.1 INTRODUCTION

The trend of schooling is changing over time. School education is also getting embedded through new methods of teaching and learning activities, for example nowadays many schools are adopting smart class technology, practical demonstrations of topics and debates between classmates. These methods are encouraged by parents. But, this smart learning method is yet to be accepted and improvised.

With this aim in mind, data mining forms the basis for searching new patterns in data. When data is stored in a database or data warehouse, then database designers can develop relational associations of data elements. When designing the database, that data contains intentionally more information and more relationships. Different algorithms and techniques are used in data mining, including visualization techniques and simple counting as well as more complex algorithms such as clustering, rule-based, fuzzy logic, decision trees and neural networks.

Email: apurva.vashist@gmail.com; suchismita.mishra8@gmail.com

DOI: 10.1201/9781003216278-9

With swift developments in the education system, several new teaching–learning methods along with many innovative and creative technologies have been deployed to improve traditional teaching methods. A new trend is found in both e-learning and classroom teaching methods – the collaborative learning method, which has become the dominant method. Based upon the preferences of modern society, in education methods using a collaborative learning technique have redefined the teacher–student relationship. In this trend of collaborative learning a number of students complete their task which is assigned by their teacher; they study together to solve the problem or successfully complete the assignment. In this method students perform better than with individual learning because they learn while discussing and sharing their ideas and thoughts with each other.

9.2 LITERATURE REVIEW

Clustering is used to help evaluate learning by Agathe Merceron. In the current research paper we attempted to show how clustering techniques can be implemented in student answers. Specifically our interest was in extracting all the clusters of students based on mistakes they had made with the aim of obtaining pedagogically relevant information and then providing this feedback to the respective subject teachers. The data used here was generated from Logic-ITA, a web-based tutoring tool. This tool is practiced for formal proofs and used in the School of Information Technologies, University of Sydney.

Blending learning has become the focus of an increasing amount of research. Liu, Ba, Huang Wu and Lao (2017) studied collaborative learning and indicate that it is an important teaching strategy for learning performance, either in e-learning or classroom teaching. Different students have different learning styles and it is difficult for teachers to take these factors into consideration.

In data mining and inductive learning, the decision tree plays an important role. C4.5 is a classic classification algorithm used in data mining. Its efficiency significantly decreases when it is used in mass calculations. In a paper by Rong Cao and Xu (2009), C4.5 algorithm efficiency is improved with the use of L'Hospital rules. The method simplifies the calculation process to improve C4.5 efficiency used in the decision-making process. The application described in this paper demonstrates that the new algorithm is more efficient and suitable for application on a larger data set.

In a paper by Rui Li et al. (2009), a balanced coefficient is introduced to improve the veracity factor in the C4.5 algorithm. This harmonizes the implementation guide (IG) ratio of each of the attributes with an artificial method in particular surroundings. From the improved algorithm, classification results have much more veracity and it enables a rational decision tree (Rui Li, Xian-mei Wei and Xue-wei Yu, 2009).

The decision tree is a technique that converts fact into a tree structure representing comprehensible rules. This decision tree is beneficial for exploring a data set, finding hidden relationships between input data and targeted variables. The main purpose of a study by Sri Wahyuni et al. (2017) was to classify a student database at the University of Pembangunan Panca Budi to find out factors influencing why students drop out.

The attributes studied were school of origin, age, occupation of parents, parents' income and their grade point score. To avoid overbranching of the tree, the factors income, age and score were grouped into a single attribute. It was found that the most influential factor in a student dropping out was school of origin. The calculated result found an accuracy value of 59.58%. The classification error found was only 40.42%.

A paper by Agrawal and Gupta (2013) proposed the C4.5 algorithm, improved by implementation of L'Hospital rule to simplify calculations and improve the efficiency of the algorithm. Here the system aims to implement this modified algorithm globally to compare its complexities and efficiency.

9.3 CLUSTER ANALYSIS

In cluster analysis several data objects are divided into clusters. In the same cluster all data objects are similar to each other although different from another cluster. Cluster analysis is used in many areas, for example in business and market research. By using these analyses we can segregate customers into different groups according to their interests and shopping habits.

Cluster analysis itself is not an algorithm. It is completed through different algorithms according to what kind of clustering is required. Generally clustering algorithms may be classified into four categories: exclusive, overlapping, hierarchical and probabilistic clustering.

The C4.5 algorithm is arguably one of the most famous techniques popularly implemented in data mining as the decision tree classifier is known to be employed to generate a perfect decision, based on certain samples of data. This works well with both univariate and multivariate predictors.

9.4 PROPOSED MODEL

This study was carried out in four developing small townships of north India. It includes parameters like: gender, distance from school, faculty qualification, infrastructure, fee structure and number of siblings.

From Figure 9.1, it can be seen that one of the prime factors influencing the decision is parents visiting. This opens up an array of conditions. Now, if the number of children in a family is one, two or more than two, the parents either decide on primary school or upper primary school, respectively. But, if the number of children is two or more than two then it depends on how they make the decision on how to give schooling to both.

The root of the tree is usually the variable which has the minimum value in a cost function. In this particular example, the probability of parents visiting is 50% each, leading to easier decision making. But what if no of children were selected as the root? Then there would be a 33.33% chance of each happening, which increases our chances of making a wrong decision because there are more test cases to consider. This is understandable if we go through the concepts of information gain (IG) and entropy.

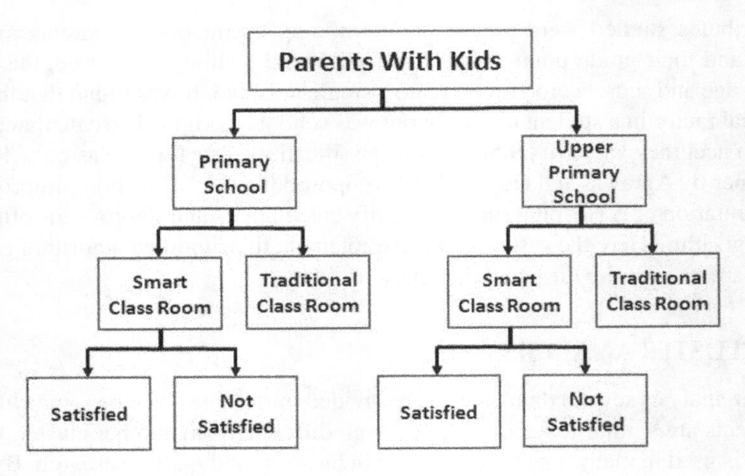

FIGURE 9.1 Decision tree depicting parents' preference for a smart school.

9.5 INFORMATION GAIN

Whether something is going to happen or not can be predicted accurately and precisely with the information acquired over time; in that case the information collected regarding that specific event which is predicted is not new information. However, if the situation reverses and there are unexpected occurrences for any reason, this may be considered useful necessary information. A similar concept is IG.

From the above statement, we can formulate that the amount of information gained is indirectly proportional to the probability of an event occurring. In addition, we can say that as the entropy increases the IG reduces, as entropy refers to the probability of an event.

Let's say, we are looking at the example of the toss of a coin. The probability of either side of a fair coin coming on top is 50%. If the coin is unfair or biased such that the probability of obtaining either a head or tail on top is 1.00, then we say that the entropy is minimum as without any trials we can predict the outcome of the coin toss.

$$H(X) = -\sum_{i=1}^{n} P(x_i)\log_b P(x_i) = -\sum_{i=1}^{2}(1/2)\log_2(1/2) = -\sum_{i=1}^{2}(1/2)\times(-1) = 1 \quad (9.1)$$

From Figure 9.2, we notice that the maximum information gained due to maximized uncertainty of a particular event is when the probability of all of the events is exactly equal.

In this case, $p = q = 0.5$ where Σ is the entropy of the system event, p is the probability of 'head' on the top as outcome, and q is the probability of 'tail' on the top as outcome.

In the case of decision trees, it is important that the nodes are aligned so that entropy reduces with splitting downwards. This indicates that as splitting is done properly, coming to an exact decision becomes relatively easier.

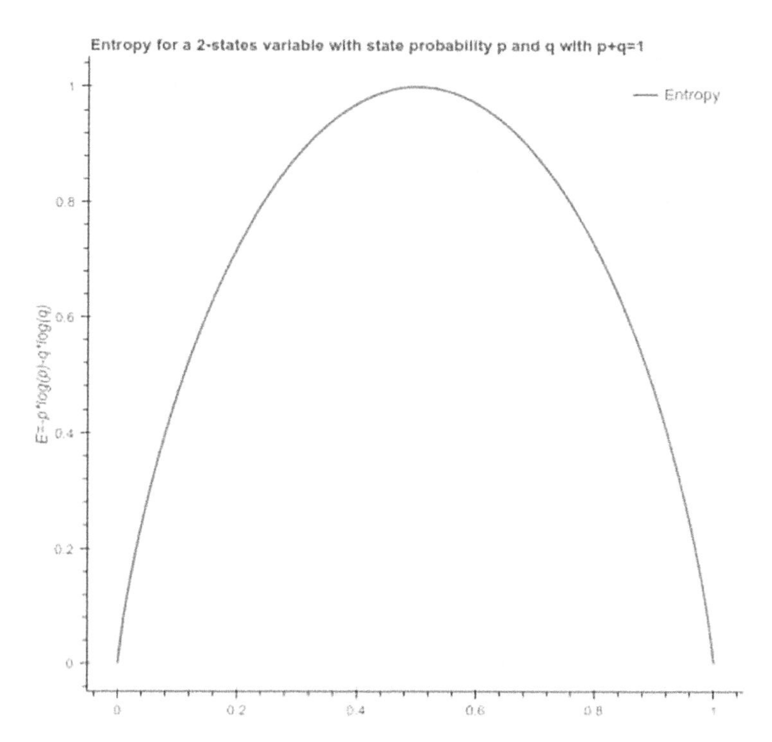

FIGURE 9.2 Probability of the decision process.

So, we check every node against every possible splitting possibility. The IG ratio is defined as the ratio of observations to the total number of observations:

$$(m/N = p) \text{ and } (n/N = q) \text{ and in which } m + n = Nm + n = N \text{ and } p + q = 1p + q = 1 \quad (9.2)$$

In the example, we find that the parents visiting reduces entropy to a greater scale compared to other options. Hence, we proceed with this option here.

9.6 EXPERIMENTAL SETUP

This study was done in two steps: (1) exploratory phase and (2) measurement phase. The noteworthy factors were determined by applying principal component analysis. The C4.5 algorithm was applied to the resulting data to determine the co-relation factors.

In the exploratory phase of the study, data was collected from trade partners on a pen-and-paper basis describing their satisfaction level with each touch point. The touch points are:

- Gender:
  - Male
  - Female
- Number of children:
  - One
  - Two
  - Three
  - More than three
- Parent with children:
  - Going to primary classes
  - Going to upper primary classes

This reveals parents' expenditure limits and their trust in the school. Also the faculty's qualifications play an important role, because parents prefer highly qualified over averagely qualified faculties and believe that highly qualified faculties will be more beneficial for their child.

9.6.1 Pseudo-Code

1. As a first step, check the base cases (directly above).
2. Find the normalized form of the IG ratio by splitting on the '*a*', for each attribute *a*.
3. Let us assume the best '*a*'_ is the attribute with the highest normalized IG value.
4. Now we can generate a decision node of the tree which splits on the a_ best value.
5. Repeat the sub-lists which are attained from splitting of the a_best value, and from adding those nodes as its child node.

9.6.2 Benefits of C4.5 Compared with Other Existing Decision Tree Systems

1. This algorithm intrinsically works the single-pass tree node pruning method to diminish over-fitting.
2. This method performs well with and without both discrete and continuous data sets.
3. The C4.5 algorithm is able to handle efficiently those issues that are generated due to an incomplete data set.

9.7 RESULTS AND ANALYSIS

From the above study of the clusters the visualizations given below were obtained. The study was carried out in four different small townships of north India on the responses and preferences of parents regarding their children's education. In the study the parameters that were mentioned by most parents were:

- Popularity of the school
- Good and safe bus service
- Reputation of the principal
- Good faculty

Using these parameters, 500 parents were studied; it was found that depending on safety, quality of faculty and infrastructure, parents were also willing to send their children to satellite schools (Figures 9.3–9.7).

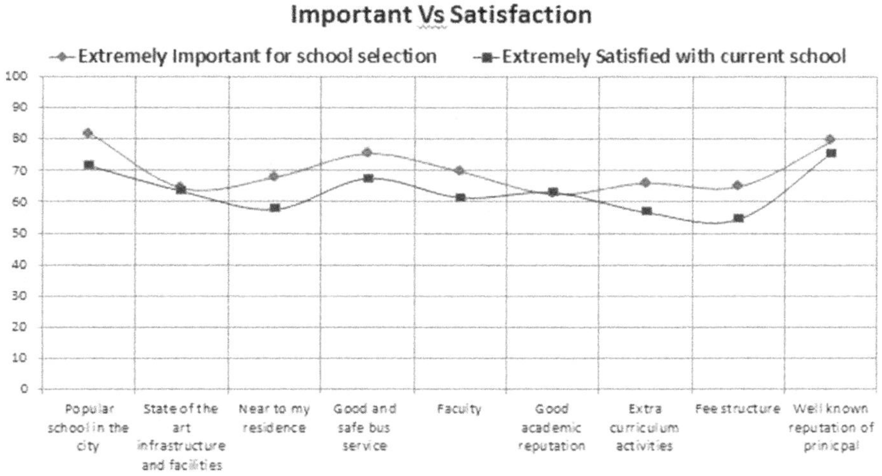

FIGURE 9.3 The overall importance of a good school vs. satisfaction with the current school.

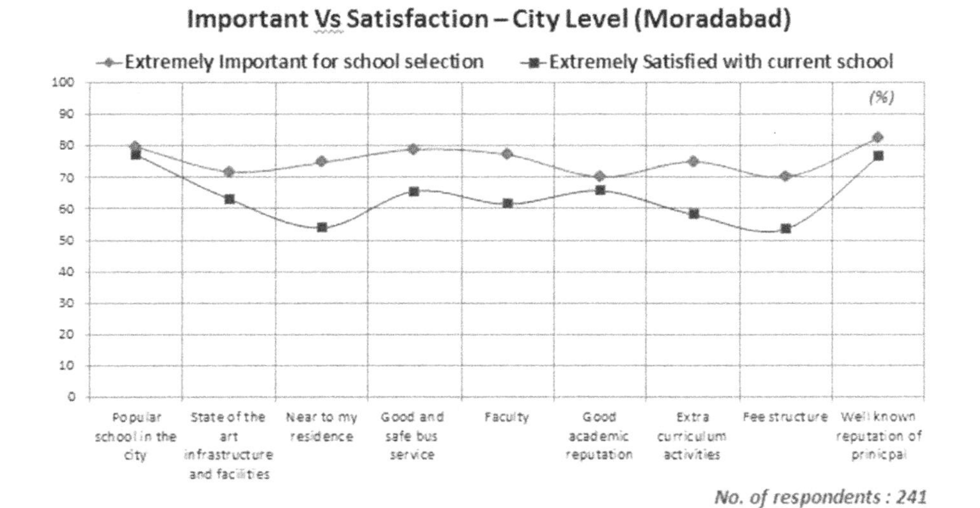

FIGURE 9.4 Acceptance response of a new school over an existing school in Moradabad.

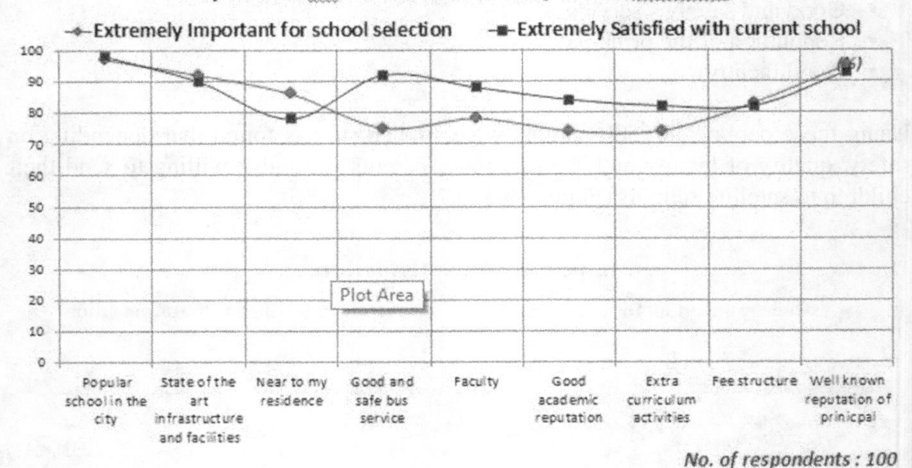

FIGURE 9.5 Acceptance response of a new school over an existing school in Sambal.

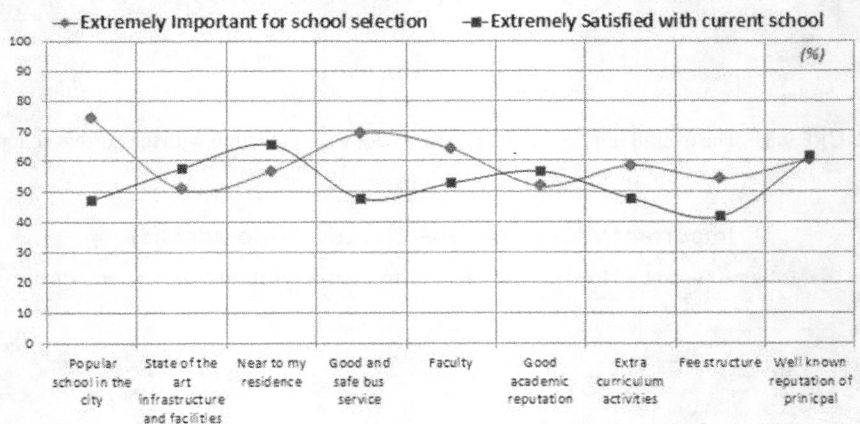

FIGURE 9.6 Acceptance response of a new school over an existing school in Amroha.

From the above study it can clearly be observed that the infrastructure of the school is nowadays a greater priority when parents are choosing a school for their children. Also, a safe and good environment plays an important role when making decisions on choice of school. So it is becoming an important factor for the schools also to improve their teaching–learning method as well as to adopt new methods.

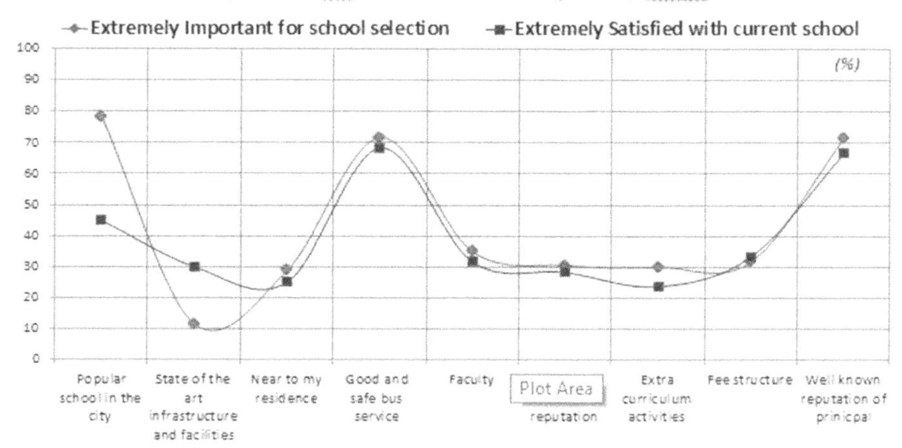

FIGURE 9.7 Acceptance response of a new school over an existing school in Joya.

9.8 CONCLUSION AND FUTURE SCOPE

Using cluster analysis, we studied different patterns of education such as blended course or online classes. By different analysis, we have concluded that the acceptance response of a new school is good over an existing school. In future, we online exams and competitions could be held, so that students can do different types of work at the same time.

BIBLIOGRAPHY

Agrawal, G.L., Gupta, H. Optimization of C4.5 decision tree algorithm for data mining application. International Journal of Emerging Technology and Advanced Engineering, 2013;3(3).

Chang, Y.C., Kao, W.Y., Chu, C.P., Chiu, C.H. A learning style classification mechanism for e-learning. Computer Education, 2009;53:273–285.

Liu, B, Huang, W., Lao. A study on grouping strategy of collaborative learning based on clustering algorithm (2017).

Liu, Q., Ba, S., Huang, J., Wu, L., Lao C. A study on grouping strategy of collaborative learning based on clustering algorithm. In: Cheung, S., Kwok, L., Ma, W., Lee, L.K., Yang, H. (eds.) Blended Learning. New Challenges and Innovative Practices. ICBL 2017. Lecture Notes in Computer Science, vol. 10309. Cham: Springer, 2017.

Ma, Y.Y., Yuan, J. Based on GSDBK – means grouping algorithm research for networked collaborative learning. Electronic Science Technology, 2016;29(12):89–92.

Mehta, M., Agrawal, R., Rissanen, J. SLIQ: a fast scalable classifier for data mining. In: Apers, P., Bouzeghoub, M., Gardarin, G. (eds.) Proceedings of the 5th International Conference on Extending Database Technology. Berlin: Springer-Verlag, 1996, pp. 18–32.

Maina, M.E., Oboko, O.R., Waiganjo, W.P. Using machine learning techniques to support group formation in an online collaborative learning environment. Journal of Intelligent Systems and Applications 2017;3:26–33.

McLaren, B.M., Scheuer, O., Miksatko, J. Supporting collaborative learning and e-discussions using artificial intelligence techniques. International Journal of Artificial Intelligence Education 2010;20(1):1–46.

Merceron, A., Yacef, K. Clustering Students to Help Evaluate Learning. 2004. https://telearn.archives-ouvertes.fr/ha-l00190274

Rong, C., Lizhen, X. Improved C4.5 algorithm for the analysis of sales. In: Sixth Web Information Systems and Applications Conference. IEEE, 2009.

Rui, L., Xian-mei W., Xue-wei, Y. The improvement of C4.5 algorithm and case study. In: Second International Symposium on Computational Intelligence and Design. IEEE, 2009.

Varghes, B., Jacob, P., Tomy, J. Clustering students' data to characterize performance patterns. International Journal of Advanced Computer Science and Application, special issue, 2011.

Wahyuni, S., Saputra, K., Mochammad, I.P.-A. The implementation of decision tree algorithm C4.5 using RapidMiner in analyzing dropout students. In: Fourth International Conference on Technical and Vocation Education and Training, Padang, November 9–11, 2017.

Webber, G., Lima, P. Evaluating automatic group formation mechanisms to promote collaborative learning: a case study. International Journal of Learning Technology, 2012;7(3).

10 Two-Phase Natural Convection of Dusty Fluid Boundary Layer Flow over a Vertical Plate

Sasanka Sekhar Bishoyi and Ritesh Dash

10.1 INTRODUCTION

Two-phase free convection has been occurring widely in the environment for many years due to its vast importance in nature and its applications. The investigation of heat exchanges in buoyancy-influenced flows is important for a number of applications in engineering, including cooling of electronics, heating and cooling of buildings, heat transfer processes and safety applications. The need for natural convection along a flat plate has been studied due to its vast applications in the field of engineering like cooling of electronics equipment and in the crystal growth process. It remains a subject of interest either experimentally or theoretically due to the number of possible variations in boundary conditions. In recent years, assorted endeavors have been undertaken, delving into the problem of free convection over a vertical wall in a laminated medium due to its clear dominance. In the past researchers were fascinated to search out solutions due to the similar variables which can give great physical insight with minimum effort.

Yang [1] has developed an omnibus outlook for prevailing solutions to a class of problems for a non-isothermal vertical wall surrounded by an isothermal atmosphere. Sparrow, Eichhorn and Grigg [2] analyzed numerical solutions of equations for a non-isothermal flat plate which is vertical for two families of wall temperature for various values of the Prandtl number using a transformation involving the Grashof

Email: ssbishoyi@gmail.com; rdasheee@gmail.com

DOI: 10.1201/9781003216278-10

number. Little attention has been paid to free convection flow with suspended particulate matter.

Many authors have reviewed multiphase flow equations. Soo [3] has come up with a numerical outlook to such type of flow. Rietema and Van der Akker [4] addressed the comprehensive derivation of momentum equations for disseminated two-phase systems. Saffmann [5] addressed the stability of dusty gas laminar flow in which the dust particles are distributed uniformly. Asmolov [6] examined the laminar flow of dusty fluid over a curved surface. Palani and Ganesan [7] studied the effect of heat transfer and velocity profiles on dusty gas flow past a semi-infinite inclined plate. Abbassi et al. [8] considered the case of natural convection from elliptic cross-section cylinders.

The main objective of this study was to analyze by natural convection for both phases, namely liquid and dust particles, in heat transfer and velocity distribution. Here, the effects of buoyancy force and dimensionless parameters like volume fraction (φ), Nusselt number (Nu), Prandtl number (pr), concentration parameter (α) , and diffusion parameter (ϵ) on skin friction and heat transfer were studied. The non-linear acquired system of equations for both phases have been solved numerically by an implicit finite difference method.

10.2 MATHEMATICAL FORMULATION

In this chapter we have considered an extremely long and wide free convection flow in the plane which is perpendicular to the floor. The x and y coordinates were chosen along and perpendicular to the plate. Further surface temperature of the heated or cooled layer was taken as T_w [9, 10].

A boundary layer approximation for particle phase momentum equations is not necessary and also momentum equations for particles cannot be neglected in the transverse direction [11, 12]. Under the above circumstances, the y component for the fluid phase of the momentum equation is dropped whereas the y component for the particle phase of the momentum equation is retained [13].

The two-phase boundary layer equations considering the above suppositions are:

$$\frac{\partial u}{\partial x}+\frac{\partial v}{\partial y}=0 \tag{10.1}$$

$$u_p\frac{\partial \rho_p}{\partial x}+v_p\frac{\partial \rho_p}{\partial y}=v_p\frac{\partial^2 \rho_p}{\partial y^2} \tag{10.2}$$

$$(1-\varphi)\rho\left(u\frac{\partial u}{\partial x}+v\frac{\partial u}{\partial y}\right)=(1-\varphi)\mu\frac{\partial^2 u}{\partial y^2}-\frac{1}{\tau_p}\varphi\rho_s\left(u-u_p\right)+(1-\varphi)\rho g\beta\left(T-T_\infty\right) \tag{10.3}$$

$$\varphi\rho_s\left(u_p\frac{\partial u_p}{\partial x}+v_p\frac{\partial u_p}{\partial y}\right)=\frac{\partial}{\partial y}\left(\varphi\mu_s\frac{\partial u_p}{\partial y}\right)+\frac{1}{\tau_p}\varphi\rho_s\left(u-u_p\right)+\varphi\left(\rho_s-\rho\right)g \tag{10.4}$$

$$\varphi\rho_s\left(u_p\frac{\partial v_p}{\partial x}+v_p\frac{\partial v_p}{\partial y}\right)=\frac{\partial}{\partial y}\left(\varphi\mu_s\frac{\partial v_p}{\partial y}\right)+\frac{1}{\tau_p}\varphi\rho_s\left(v-v_p\right) \tag{10.5}$$

$$(1-\varphi)\rho c_p\left(u\frac{\partial T}{\partial x}+v\frac{\partial T}{\partial y}\right)=(1-\varphi)k\frac{\partial^2 T}{\partial y^2}+\frac{1}{\tau_T}\varphi\rho_s c_s\left(T_p-T\right)+(1-\varphi)\mu\left(\frac{\partial u}{\partial y}\right)^2 \tag{10.6}$$

$$\begin{aligned}\varphi\rho_s c_s\left(u_p\frac{\partial T_p}{\partial x}+v_p\frac{\partial T_p}{\partial y}\right)&=\frac{\partial}{\partial y}\left(\varphi k_s\frac{\partial T_p}{\partial y}\right)-\frac{1}{\tau_T}\varphi\rho_s c_s\left(T_p-T\right)\\&+\varphi\mu_s\left[u_p\frac{\partial^2 u_p}{\partial y^2}+\left(\frac{\partial u_p}{\partial y}\right)^2\right]\end{aligned} \tag{10.7}$$

If temperature deviation is insignificant; D_p, μ_s and k_s may be taken as constant [14, 15]. Here, the term $\frac{\partial}{\partial y}\left(\varphi\mu_s\frac{\partial u_p}{\partial y}\right)$ may be succeeded by $\varphi\mu_s\frac{\partial^2 u_p}{\partial y^2}$; in an energy equation for the particle phase the term $\frac{\partial}{\partial y}\left(\varphi k_s\frac{\partial T_p}{\partial y}\right)$ may be replaced by $\varphi k_s\frac{\partial^2 T_p}{\partial y^2}$. In general, $D_p \approx \upsilon_p$ and D_p are often very small in comparison with υ.

By considering the above, Equations (10.1) to (10.7) reduce to:

$$\frac{\partial u}{\partial x}+\frac{\partial v}{\partial y}=0 \tag{10.8}$$

$$u_p\frac{\partial \rho_p}{\partial x}+v_p\frac{\partial \rho_p}{\partial y}=v_p\frac{\partial^2 \rho_p}{\partial y^2} \tag{10.9}$$

$$u\frac{\partial u}{\partial x}+v\frac{\partial u}{\partial y}=v\frac{\partial^2 u}{\partial y^2}-\frac{1}{1-\varphi}\frac{1}{\tau_p}\frac{\rho_p}{\rho}\left(u-u_p\right)+g\beta\left(T-T_\infty\right) \tag{10.10}$$

$$u_p\frac{\partial u_p}{\partial x}+v_p\frac{\partial u_p}{\partial y}=v_s\frac{\partial^2 u_p}{\partial y^2}+\frac{1}{\tau_p}\left(u-u_p\right)+\left(1-\frac{\rho}{\rho_s}\right)g \tag{10.11}$$

$$u_p\frac{\partial v_p}{\partial x}+v_p\frac{\partial v_p}{\partial y}=v_s\frac{\partial^2 v_p}{\partial y^2}+\frac{1}{\tau_p}\left(v-v_p\right) \tag{10.12}$$

$$u\frac{\partial T}{\partial x}+v\frac{\partial T}{\partial y}=\frac{\kappa}{\rho C_p}\frac{\partial^2 T}{\partial y^2}+\frac{1}{1-\varphi}\frac{1}{\tau_T}\frac{\rho_p}{\rho}\frac{C_s}{C_p}\left(T_p-T\right)+\frac{\mu}{\rho C_p}\left(\frac{\partial u}{\partial y}\right)^2 \tag{10.13}$$

$$u_p \frac{\partial T_p}{\partial x} + v_p \frac{\partial T_p}{\partial y} = \frac{\kappa_s}{\rho_s C_s} \frac{\partial^2 T_p}{\partial y^2} - \frac{1}{\tau_T}\left(T_p - T\right) + \frac{\mu_s}{\rho_s C_s}\left[u_p \frac{\partial^2 u_p}{\partial y^2} + \left(\frac{\partial u_p}{\partial y}\right)^2\right] \tag{10.14}$$

Introducing non-dimensional variables like:

$$x^* = \frac{x}{L}, y^* = \frac{y}{L}\sqrt{Re}, u^* = \frac{u}{U}, v^* = \frac{v}{U}\sqrt{Re}, u_p^* = \frac{u_p}{U}, v_p^* = \frac{v_p}{U}\sqrt{Re}$$

$$T^* = \frac{T - T_\infty}{T_w - T_\infty}, T_p^* = \frac{T_p - T_\infty}{T_w - T_\infty}, \rho_p^* = \frac{\rho_p}{\rho_{p0}}, \rho_s^* = \frac{\rho_s}{\rho_{p0}} \tag{10.15}$$

in Equations (10.8) to (10.14) and after dropping asterisks, the boundary layer equations are:

$$\frac{\partial u}{\partial x} + \frac{\partial v}{\partial y} = 0 \tag{10.16}$$

$$u_p \frac{\partial \rho_p}{\partial x} + v_p \frac{\partial \rho_p}{\partial y} = \epsilon \frac{\partial^2 \rho_p}{\partial y^2} \tag{10.17}$$

$$u \frac{\partial u}{\partial x} + v \frac{\partial u}{\partial y} = \frac{\partial^2 u}{\partial y^2} - \alpha \frac{1}{1-\varphi} \frac{FL}{U} \rho_p \left(u - u_p\right) + \frac{GrT}{Re^2} \tag{10.18}$$

$$u_p \frac{\partial u_p}{\partial x} + v_p \frac{\partial u_p}{\partial y} = \epsilon \frac{\partial^2 u_p}{\partial y^2} + \frac{FL}{U}\left(u - u_p\right) + \frac{1}{Fr}\left(1 - \frac{1}{\gamma}\right) \tag{10.19}$$

$$u_p \frac{\partial v_p}{\partial x} + v_p \frac{\partial v_p}{\partial y} = \epsilon \frac{\partial^2 v_p}{\partial y^2} + \frac{FL}{U}\left(v - v_p\right) \tag{10.20}$$

$$u \frac{\partial T}{\partial x} + v \frac{\partial T}{\partial y} = \frac{1}{Pr} \frac{\partial^2 T}{\partial y^2} + Ec\left(\frac{\partial u}{\partial y}\right)^2 + \frac{2\alpha}{3Pr} \frac{1}{1-\varphi} \frac{FL}{U} \rho_p \left(T_p - T\right) \tag{10.21}$$

$$u_p \frac{\partial T_p}{\partial x} + v_p \frac{\partial T_p}{\partial \eta} = \frac{FL}{U}\left(T - T_p\right) + \frac{\epsilon}{Pr} \frac{\partial^2 T_p}{\partial y^2} + \frac{3}{2}.Pr.\epsilon.Ec\left[\left(\frac{\partial u_p}{\partial y}\right)^2 + u_p \frac{\partial^2 u_p}{\partial y^2}\right] \tag{10.22}$$

Subject to boundary conditions:

$$y = 0 : u = 0, v = 0, u_p = u_{pw}(x), v_p = 0, \rho_p = \rho_{pw}(x), \mathrm{T} = 1, T_p = T_{pw}(x) \quad (10.23)$$

$$y = \infty : u = u_p = 0,\ v_p = 0, \rho_p = 1, T = 0, T_p = 0 \quad (10.24)$$

10.3 METHOD OF SOLUTION

To flourish a computational algorithm with non-uniform grid, Equations (10.16) to (10.22) are replaced by finite difference expressions [16, 17] as:

$$\frac{\partial W}{\partial x} = \frac{1.5 W_j^{n+1} - 2W_j^n + 0.5 W_j^{n-1}}{\Delta x} + o\left(\Delta x^2\right) \quad (10.25)$$

$$\frac{\partial W}{\partial y} = \frac{W_{j+1}^{n+1} - \left(1 - r_y^2\right)_j^{n+1} - r_y^2 W_{j-1}^{n+1}}{r_y\left(r_y + 1\right)\Delta y} + o\left(\Delta y^2\right) \quad (10.26)$$

$$\frac{\partial^2 W}{\partial y^2} = 2\frac{W_{j+1}^{n+1} - \left(1 + r_y\right)W_j^{n+1} + r_y W_{j-1}^{n+1}}{r_y\left(r_y + 1\right)\Delta y^2} + o\left(\Delta y^2\right) \quad (10.27)$$

$$W_j^{n+1} = 2 W_j^n - W_j^{n-1} + o\left(\Delta x^2\right) \quad (10.28)$$

and:

$$y_{j+1} - y_j = r_y\left(y_j - y_{j-1}\right) = r_y \Delta y_j \quad (10.29)$$

Equations (10.16) to (10.22) reduce into difference equations, and we get Mishra and Tripathy [9, 10]:

$$v_j^{n+1} = v_{j-1}^{n+1} - \frac{1}{2}\frac{\Delta y}{\Delta x}\left[\left(1.5u_j^{n+1} - 2u_j^n + 0.5u_j^{n-1}\right) + + \left(1.5u_{j-1}^{n+1} - 2u_{j-1}^n + 0.5u_{j-1}^{n-1}\right)\right] \quad (10.30)$$

The diffusion Equation (10.17) for particle phase yields:

$$a_j u_{j-1}^{n+1} + b_j u_j^{n+1} + c_j u_{j+1}^{n+1} = d_j \quad (10.31)$$

where:

$$a_j = \frac{1}{\Delta x}\left[-p r_y - q\right]$$

$$b_j = \frac{1}{\Delta x}\left[\begin{array}{c} 1.5\left(2u_j^n - u_j^{n-1}\right) + p\left(r_y - \frac{1}{r_y}\right) \\ +q\left(1 + \frac{1}{r_y}\right) + \frac{1}{1-\varphi}\frac{FL}{U}\alpha\Delta x\left(2\rho_{pj}^{\ n} - \rho_{pj}^{\ n-1}\right) \end{array}\right]$$

$$c_j = \frac{1}{\Delta x}\left[\frac{1}{r_y}(p-q)\right]$$

$$d_j = \frac{1}{\Delta x}\left[\begin{array}{c} \left(2u_j^n - u_j^{n-1}\right)\left(2u_j^n - 0.5u_j^{n-1}\right) \\ -\frac{\varphi}{1-\varphi}\frac{FL}{U}\alpha\Delta x\left(2\rho_{pj}^{\ n} - \rho_{pj}^{\ n-1}\right)\left(-2u_{pj}^{\ n} + u_{pj}^{\ n-1}\right) + \frac{Gr\left(2T_j^n - T_j^{n-1}\right)}{Re^2}\Delta x \end{array}\right]$$

Equation (10.18) reduces into a difference equation:

$$a_j^* u_{pj-1}^{\ n+1} + b_j^* u_{pj}^{\ n+1} + c_j^* u_{pj+1}^{\ n+1} = d_j^* \tag{10.32}$$

where:

$$a_j^* = \frac{1}{\Delta x}\left[-pr_y - \epsilon q\right]$$

$$b_j^* = \frac{1}{\Delta x}\left[1.5\left(2u_{pj}^{\ n} - u_{pj}^{\ n-1}\right) + p\left(r_y - \frac{1}{r_y}\right) + \epsilon q\left(1 + \frac{1}{r_y}\right) + \frac{FL}{U}\Delta x\right]$$

$$c_j^* = \frac{1}{\Delta x}\left[\frac{1}{r_y}(p - \epsilon q)\right]$$

$$d_j^* = \frac{1}{\Delta x}\left[\left(2u_{pj}^{\ n} - u_{pj}^{\ n-1}\right)\left(2u_{pj}^{\ n} - 0.5u_{pj}^{\ n-1}\right) + \frac{FL}{U}\Delta x u_j^{n+1} + \frac{1}{Fr}\left(1 - \frac{1}{\gamma}\right)\Delta x\right]$$

Equation (10.19) reduces into a difference equation:

$$a_j^{**} v_{pj-1}^{\ n+1} + b_j^{**} v_{pj}^{\ n+1} + c_j^{**} v_{pj+1}^{\ n+1} = d_j^{**} \tag{10.33}$$

where:

$$a_j^{**} = \frac{1}{\Delta x}\left[-pr_y - \epsilon q\right]$$

$$b_j^{**} = \frac{1}{\Delta x}\left[1.5\, u_{pj}^{n+1} + p\left(r_y - \frac{1}{r_y}\right) + \epsilon q\left(1 + \frac{1}{r_y}\right) + \frac{FL}{U}\Delta x\right]$$

$$c_j^{**} = \frac{1}{\Delta x}\left[\frac{1}{r_y}(p - \epsilon q)\right]$$

$$d_j^{**} = \frac{1}{\Delta x}\left[u_{pj}^{n+1}\left(2 v_{pj}^{n} - 0.5 v_{pj}^{n-1}\right) + \frac{FL}{U}\Delta x\, v_j^{n+1}\right]$$

Equation (10.20) reduces into a difference equation:

$$a_j^{+} T_{j-1}^{n+1} + b_j^{+} T_j^{n+1} + c_j^{+} T_{j+1}^{n+1} = d_j^{+} \tag{10.34}$$

where:

$$a_j^{+} = \frac{1}{\Delta x}\left[-q\left(0.5 r_y \Delta y\, v_j^{n+1} + \frac{1}{Pr}\right)\right]$$

$$b_j^{+} = \frac{1}{\Delta x}\left[1.5 u_j^{n+1} + 0.5 q\, \Delta y\, v_j^{n+1}\left(r_y - \frac{1}{r_y}\right) + \frac{q(1 + r_y)}{Pr.r_y} + \frac{2\alpha}{3Pr}\frac{1}{1 - \varphi}\frac{FL}{U}\Delta x\, \rho_{pj}^{n+1}\right]$$

$$c_j^{+} = \frac{1}{\Delta x}\left[\frac{q}{r_y}\left(0.5 \Delta y\, v_j^{n+1} - \frac{1}{Pr}\right)\right]$$

$$d_j^{+} = \frac{1}{\Delta x}\left[\begin{array}{c}\frac{2\alpha}{3Pr}\frac{1}{1 - \varphi}\frac{FL}{U}\rho_{pj}^{n+1}\left(2T_{pj}^{n} - T_{pj}^{n-1}\right)\Delta x + \Delta x.Ec\left(\frac{u_{j+1}^{n+1} - u_j^{n+1}}{\Delta y}\right)^2 \\ + u_j^{n+1}\left(2T_j^{n} - 0.5 T_j^{n-1}\right)\end{array}\right]$$

Equation (10.21) reduces into a difference equation:

$$a_j^{++} T_{pj-1}^{n+1} + b_j^{++} T_{pj}^{n+1} + c_j^{++} T_{pj+1}^{n+1} = d_j^{++} \tag{10.35}$$

where:

$$a_j^{++} = \frac{1}{\Delta x}\left[-q\left(0.5 r_y \Delta y\, v_{pj}^{n+1} + \frac{\epsilon}{Pr}\right)\right]$$

$$b_j^{++} = \frac{1}{\Delta x}\left[1.5u_{pj}^{\;n+1} + 0.5q\Delta y\, v_{pj}^{\;n+1}\left(r_y - \frac{1}{r_y}\right) + \frac{\epsilon\, q\left(1+r_y\right)}{Pr.r_y} + \frac{FL}{U}\Delta x\right]$$

$$c_j^{++} = \frac{1}{\Delta x}\left[\frac{q}{r_y}\left(0.5\Delta y.v_{pj}^{\;n+1} - \frac{\epsilon}{Pr}\right)\right]$$

$$d_j^{++} = \frac{1}{\Delta x}\left[\begin{array}{c} u_{pj}^{\;n+1}\left(2T_{pj}^{\;n} - 0.5T_{pj}^{\;n-1}\right) + \dfrac{FL}{U}T_j^{n+1}\Delta x \\ +\dfrac{3}{2}Pr.\epsilon.Ec.\Delta x\left\{\left(\dfrac{u_{pj+1}^{\;n+1} - u_{pj}^{\;n+1}}{\Delta y}\right)^2 + 2u_{pj}^{\;n+1}\left(\dfrac{u_{pj-1}^{\;n+1} - \left(1+\dfrac{1}{r_y}\right)u_{pj}^{\;n+1} + \dfrac{1}{r_y}u_{pj+1}^{\;n+1}}{\left(1+r_y\right)\Delta y^2}\right)\right\}\end{array}\right]$$

Equation (10.22) reduces into a difference equation:

$$a_j\rho_{pj-1}^{\;n+1} + b_j\rho_{pj}^{\;n+1} + c_j\rho_{pj+1}^{\;n+1} = d_j \tag{10.36}$$

where:

$$a_j = -v_{pj}^{\;n+1}r_y^2\,\Delta y - 2\epsilon\, r_y$$

$$b_j = \frac{1.5p\;u_{pj}^{\;n+1}}{\Delta x} - v_{pj}^{\;n+1}\left(1 - r_y^2\right)\Delta y + 2\epsilon\left(1+r_y\right)$$

$$c_j = p\;\; v_{pj}^{\;n+1}\Delta y - 2\epsilon$$

$$d_j = p\;\; u_{pj}^{\;n+1}\,\frac{2\rho_{pj}^{\;n} - 0.5\rho_{pj}^{\;n-1}}{\Delta x}$$

$$p = \left(2\,v_j^n - v_j^{n-1}\right)\frac{\Delta x}{\left(1+r_y\right)\Delta y}$$

$$q = \frac{2\,\Delta x}{\left(1+r_y\right)\Delta y^2}$$

$$p = r_y\left(1+r_y\right)\Delta y^2$$

10.4 RESULTS AND DISCUSSION

Numerical outcomes were carried out by considering $\rho = 0.913\ \text{kg/m}^3$; $\rho_p = 800, 2403, 8010\ \text{kg/m}^3$; $D = 50, 100$ ¼m; $\alpha = 0.1$, $U = 0.45\ \text{m/s}$; $\varphi = 0.001, 0.0003, 0.0001$; $\epsilon = 0.05, 0.1, 0.2$; $L = 0.3048\ \text{m}$ $\text{Ec} = 0.1$; $Pr = 0.71, 1.0, 7.0$; $= 1.5415 \times 10^{-5}\ \text{kg/m s}$. It has already been seen that the Grashof number Gr is much greater than the Reynolds number ($Gr \gg Re^2$), concerning free convection. A computational algorithm with non-uniform grid was employed to obtain the numerical solution, in which the scheme has been coded in FORTRAN [9]. Numerical results have been presented through figures and tables for various values of the Prandtl number (Pr), the diffusion parameter (ϵ), density of the particle (ρ_{sp}), diameter of the particle (D), volume fraction (φ), and concentration parameter (α). The temperature and velocity fields for different φ are depicted in Figures 10.1–10.4. From Figure 10.2, we can see that the effect of a greater number of particles in the unit volume of the mixture is that the magnitude of the carrier fluid and particle phase velocity are decreased.

Figures 10.5 and 10.6 depict the temperature and velocity profiles of carrier fluid whereas Figures 10.7 and 10.8 depict the temperature and velocity of particle phases for various particle material densities. It can be seen that the velocity and temperature of the carrier fluid and particle phase velocity are not significantly affected but the

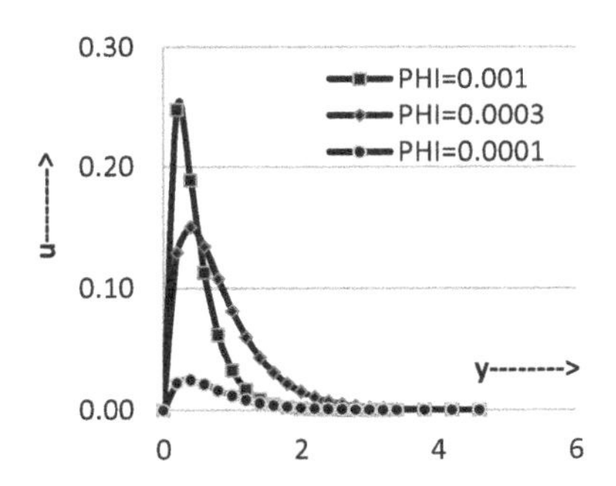

FIGURE 10.1 Alteration of u with y.

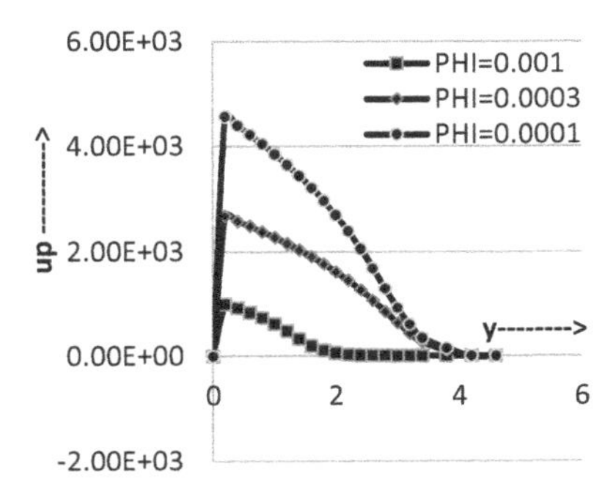

FIGURE 10.2 Alteration of U_p with y.

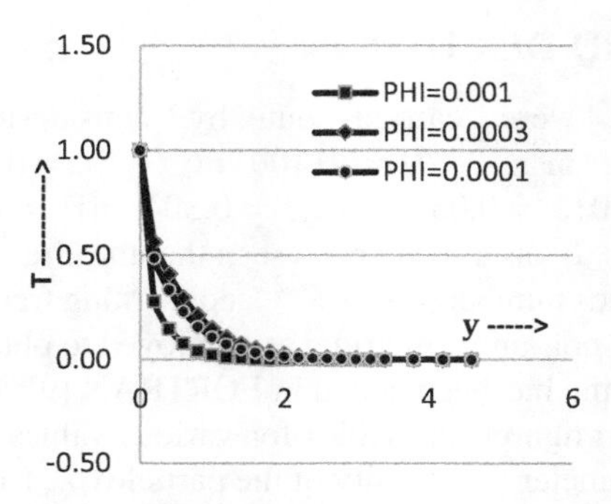

FIGURE 10.3 Alteration of T with y.

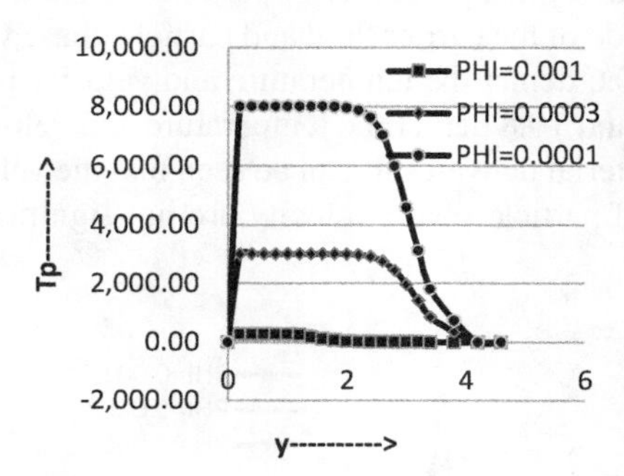

FIGURE 10.4 Alteration of T_p with y.

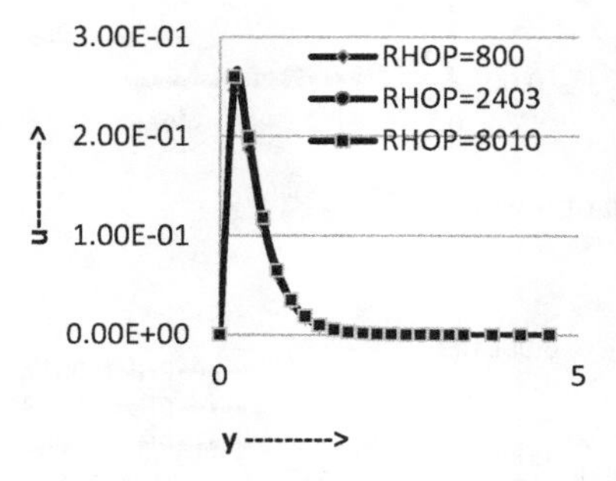

FIGURE 10.5 Alteration of u with y.

particle phase temperature increases with increase of particle material density. The alteration of temperature and velocity distribution for different values of diffusion parameter (ϵ) are illustrated in Figures 10.9–10.12. From Figures 10.9 and 10.10, it can be seen that an increase in diffusion parameter does not affect the velocity of carrier fluid and temperature, but Figures 10.11 and 10.12 show that an increase the

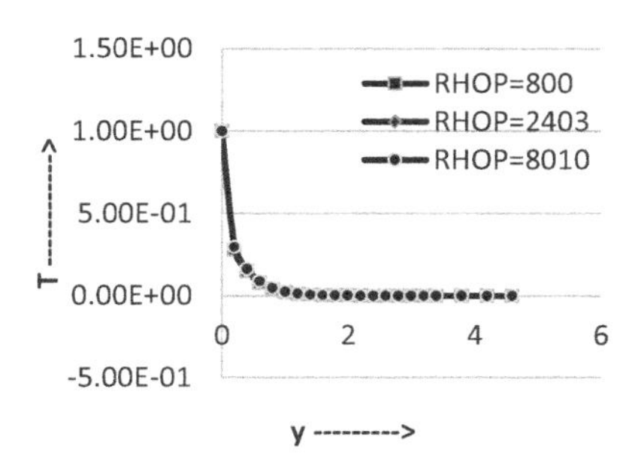

FIGURE 10.6 Alteration of T with y.

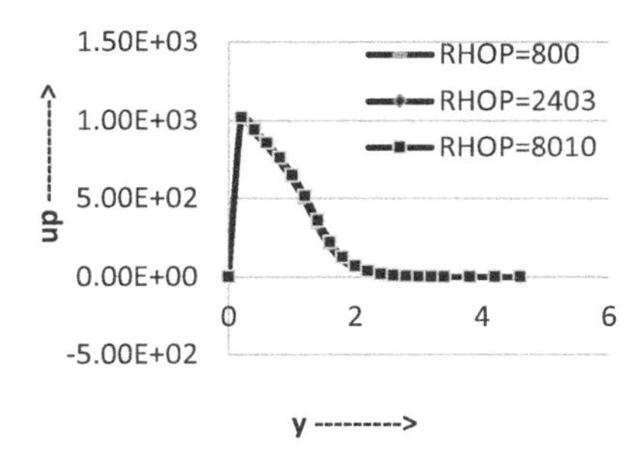

FIGURE 10.7 Alteration of U_p with y.

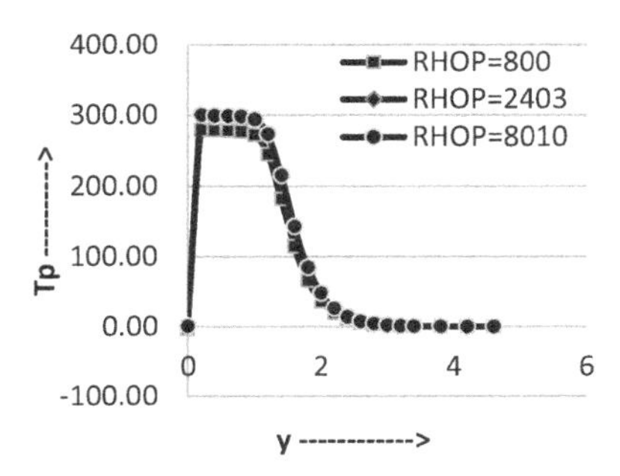

FIGURE 10.8 Alteration of T_p with y.

diffusion parameter helps to reduce in the magnitude of the particle phase temperature and velocity.

The deviation of temperature and velocity distribution for different values of p_r with y are shown in Figures 10.13–10.16. The increase in values of p_r does not affect the velocity of the carrier fluid as well as particle phase in Figures 10.13 and 10.14. From Figures 10.15 and 10.16, we observe that the increase in value of p_r decreases the temperature of the carrier fluid whereas the temperature of the particle phase increases with p_r. It is observed that if the Prandtl number increases then temperature

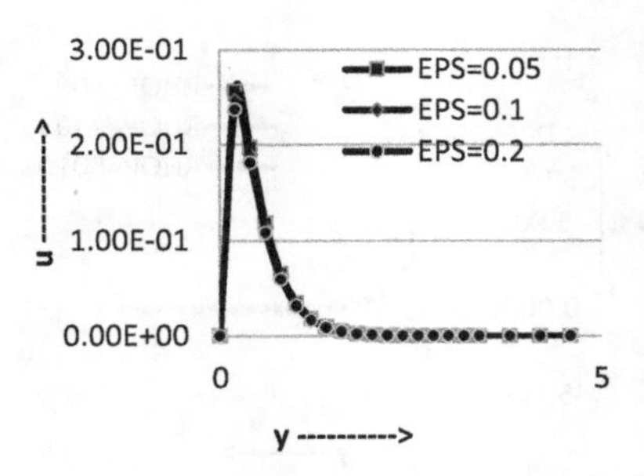

FIGURE 10.9 Alteration of *u* with *y*.

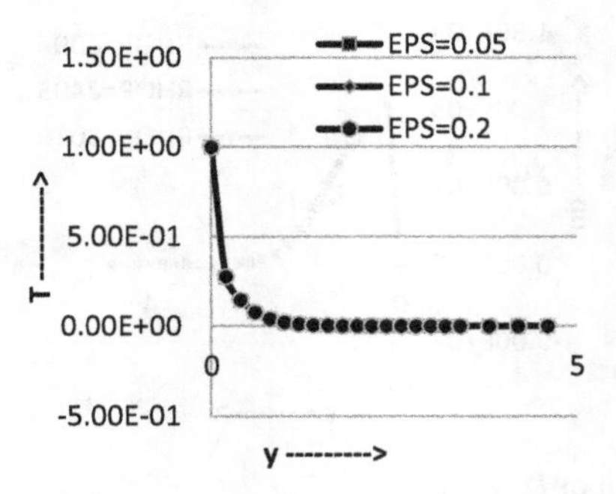

FIGURE 10.10 Alteration of *T* with *y*.

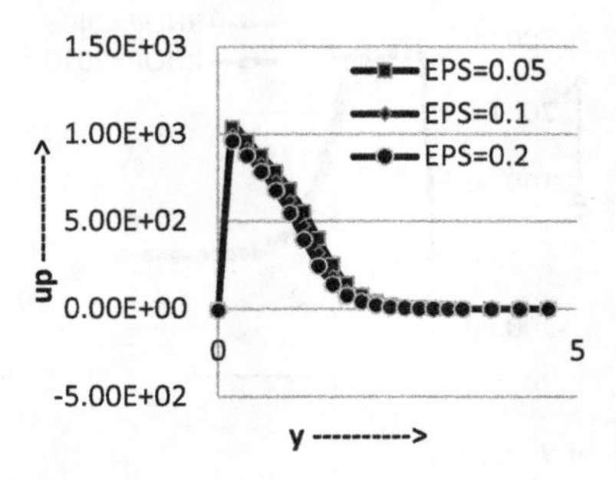

FIGURE 10.11 Alteration of U_p with *y*.

distribution decreases. Therefore, air temperature falls more rapidly than water and electrolyte as carrier fluid. Coarser particles help to transfer heat from plate to fluid, reduce skin friction, and decrease displacement thickness.

Figures 10-17–10.20 show that an increase in diameter (*D*) (i.e., the size of the particle) is associated with an increase in velocity and temperature of the particle

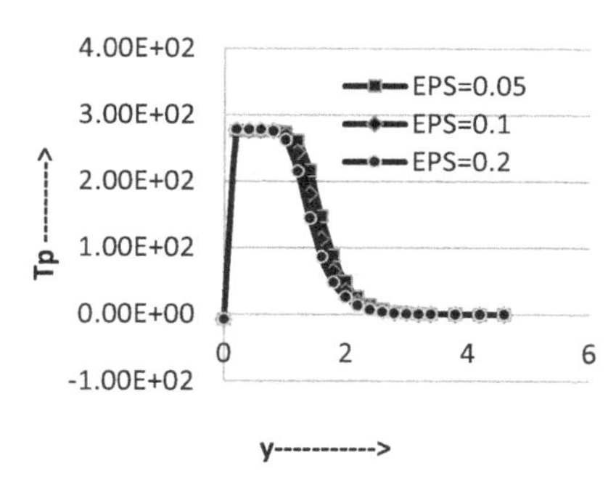

FIGURE 10.12 Alteration of T_p with y.

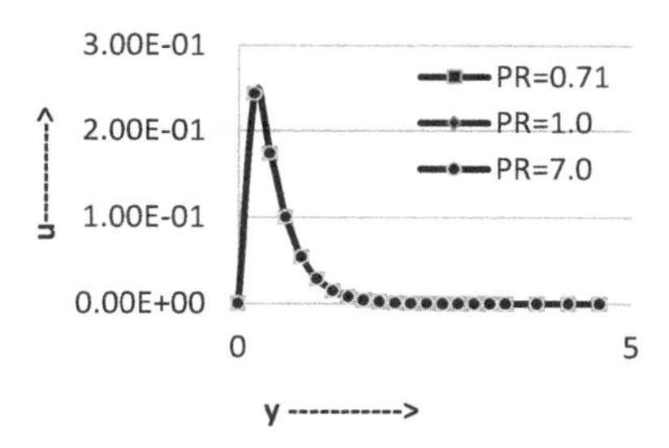

FIGURE 10.13 Alteration of u with y.

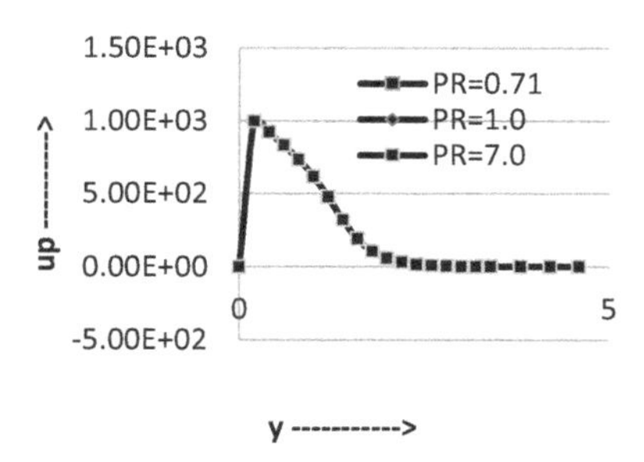

FIGURE 10.14 Alteration of U_p with y.

phase but the temperature of the carrier fluid decreases with the size of the particle. Tables 10.1–10.3 show heat transfer and skin friction heat transfer from plate to fluid with decreases in volume fraction (φ), whereas displacement thickness increases with φ. Tables 10.4–10.6 depict that the presence of coarser particles increases the heat transfer from plate to fluid, reduces skin friction, and decreases displacement thickness.

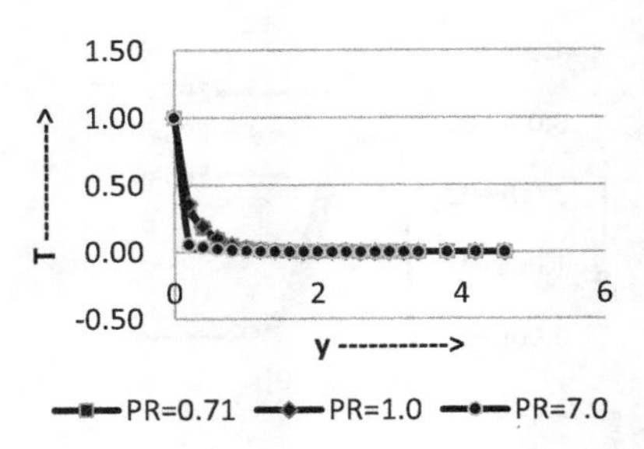

FIGURE 10.15 Alteration of T with y.

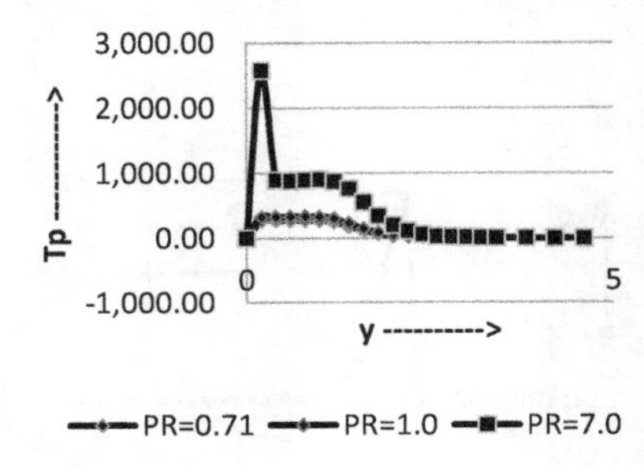

FIGURE 10.16 Alteration of T_p with y.

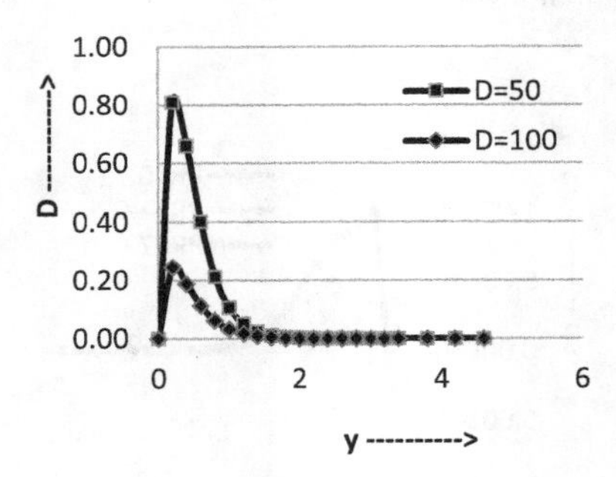

FIGURE 10.17 Alteration u with D.

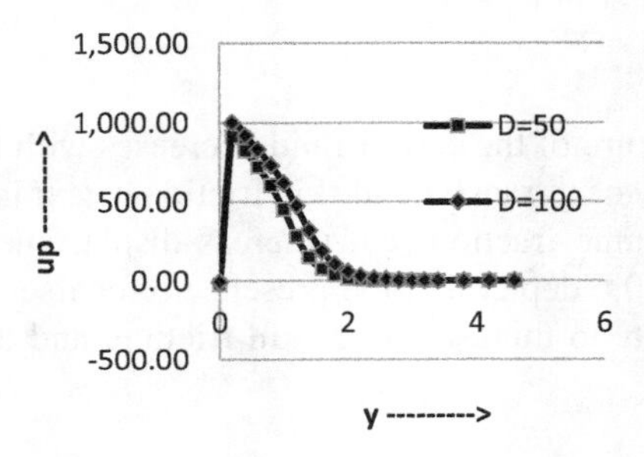

FIGURE 10.18 Alteration of U_p with D.

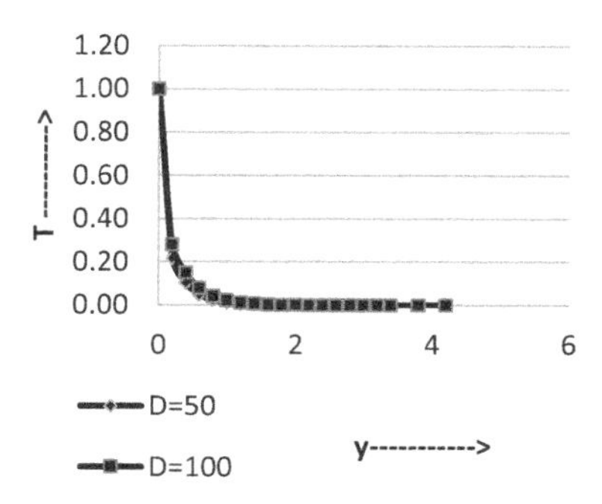

FIGURE 10.19 Alteration of T with D.

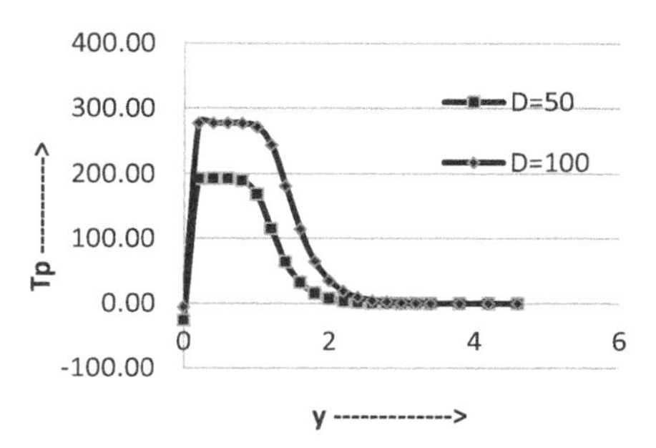

FIGURE 10.20 Alteration of T_p with D.

TABLE 10.1
Variation of Skin for Different Volume Fractions (φ)

X	$\varphi = 0.001$	$\varphi = 0.0003$	$\varphi = 0.0001$
1.20	2.70E-02	4.29E-03	2.11E-03
1.60	1.43E-02	2.14E-03	4.45E-04
2.00	4.59E-03	3.18E-03	5.97E-04
2.40	7.69E-03	2.72E-03	4.45E-04
2.80	7.64E-03	2.86E-03	5.01E-04
3.20	7.64E-03	2.93E-03	5.23E-04
3.60	7.64E-03	2.93E-03	5.22E-04
4.00	7.64E-03	2.93E-03	5.22E-04
4.40	7.64E-03	2.93E-03	5.22E-04
4.80	7.64E-03	2.93E-03	5.22E-04
5.00	7.64E-03	2.93E-03	5.22E-04

TABLE 10.2
Variation of Displacement for Different Volume Fractions (φ)

X	$\varphi = 0.001$	$\varphi = 0.0003$	$\varphi = 0.0001$
1.20	1.01E-05	3.52E-03	3.68E-03
1.60	1.98E-03	3.78E-03	3.89E-03
2.00	2.34E-03	3.73E-03	3.88E-03
2.40	2.27E-03	3.81E-03	3.89E-03
2.80	2.26E-03	3.80E-03	3.89E-03
3.20	2.26E-03	3.81E-03	3.89E-03
3.60	2.26E-03	3.81E-03	3.89E-03
4.00	2.26E-03	3.81E-03	3.89E-03
4.40	2.26E-03	3.81E-03	3.89E-03
4.80	2.26E-03	3.81E-03	3.89E-03
5.00	2.26E-03	3.81E-03	3.89E-03

TABLE 10.3
Variation of Nusselt Number for Different Values of Volume Fractions (φ)

X	$\varphi = 0.001$	$\varphi = 0.0003$	$\varphi = 0.0001$
1.20	7.44E+07	2.49E+05	4.04E+05
1.60	8.58E+07	2.43E+05	4.21E+05
2.00	5.90E+07	2.32E+05	4.14E+05
2.40	8.10E+07	2.08E+05	4.08E+05
2.80	8.10E+07	2.11E+05	4.08E+05
3.20	8.10E+07	2.13E+05	4.09E+05
3.60	8.10E+07	2.13E+05	4.09E+05
4.00	8.10E+07	2.12E+05	4.09E+05
4.40	8.10E+07	2.12E+05	4.09E+05
4.80	8.10E+07	2.12E+05	4.09E+05
5.00	8.10E+07	2.12E+05	4.09E+05

TABLE 10.4
Alteration of Skin Friction for Different Values of Diameter (D) of Particles

X	$D = 50$	$D = 100$
1.20	9.85E-02	2.70E-02
1.60	7.44E-02	1.43E-02
2.00	2.90E-02	4.59E-03
2.40	2.50E-02	7.69E-03
2.80	2.49E-02	7.64E-03
3.20	2.49E-02	7.64E-03
3.60	2.50E-02	7.64E-03
4.00	2.50E-02	7.64E-03
4.40	2.50E-02	7.64E-03
4.80	2.50E-02	7.64E-03
5.00	2.50E-02	7.64E-03

TABLE 10.5
Alteration of Displacement (*D*) for Different Values of Diameter of Particles

X	*D* = 50	*D* = 100
1.20	–6.14E-03	1.01E-05
1.60	–2.63E-03	1.98E-03
2.00	3.43E-04	2.34E-03
2.40	1.09E-03	2.27E-03
2.80	1.19E-03	2.26E-03
3.20	1.24E-03	2.26E-03
3.60	1.27E-03	2.26E-03
4.00	1.30E-03	2.26E-03
4.40	1.32E-03	2.26E-03
4.80	1.34E-03	2.26E-03
5.00	1.34E-03	2.26E-03

TABLE 10.6
Alteration of Nusselt Number for Different Values of Diameter (*D*) of Particles

X	*D* = 50	*D* = 100
1.20	4.69E+08	7.44E+07
1.60	1.01E+09	8.58E+07
2.00	2.90E+08	5.90E+07
2.40	5.49E+08	8.10E+07
2.80	5.50E+08	8.10E+07
3.20	5.51E+08	8.10E+07
3.60	5.52E+08	8.10E+07
4.00	5.52E+08	8.10E+07
4.40	5.52E+08	8.10E+07
4.80	5.53E+08	8.10E+07
5.00	5.53E+08	8.10E+07

REFERENCES

1. Yang K.T. [1960], "Possible similarity solution for laminar free convection vertical plates and cylinders", *J. Appl. Mech.*, 82:230–236.
2. Sparrow E.M., Eichhorn R. & Grigg J.L. [1959], "Combined forced and free convection in a boundary layer", *Physics Fluids,* 2:319–320.
3. Soo S.L. [1967], "Fluid Dynamics of Multiphase Systems", Waltham, MA: Blaisdell.
4. Rietema K. & Van der Akker H.E.A. [1983], "On the momentum equations in the dispersed two-phase system", *Int. J. Multiphase Flow*, 9(1):31–36.
5. Saffmann P.G. [1962], "On the stability of laminar flow of a dusty-gas", *J. Fluid Mechanics*, 13:120–128.
6. Asmolov E.S. [1995], "Dusty-gas flow in a laminar boundary layer over a blunt body", *J Fluid Mechanics*, 305:29.

7. Palani G. & Ganesan P. [2007], "Heat transfer effects on dusty-gas flow past a semi-infinite inclined plate", Forsch. Ingenieurwesen, 70(3–4):223.
8. Abbassi A., Basirat-Tabrizi H. & Nazemzadesh S.A. [2002], "Numerical analysis of steady-state laminar free convection over a horizontal cylinder with elliptic cross section", in S. Dost, H. Struchtrup and I. Dincer (eds.), Progress in Transport Phenomena, Elsevier, pp. 37–41.
9. Mishra S.K. & Tripathy P.K. [2011], "Mathematical and numerical modeling of two phase flow and heat transfer using non-uniform grid", *Far East J. Appl. Math.*, 54(2):107–126.
10. Mishra S.K. & Tripathy P.K. [2011], "Approximate solution of two phase thermal boundary layer flow", *Reflections des ERA*, 6(2):113–148.
11. Mahdy, A. [2019], "Entropy generation of tangent hyperbolic nanofluid flow past a stretched permeable cylinder: variable wall temperature", Proc. IMechE Part E: J. Process Mech. Engin., 233:570–580.
12. Mahdy A. & Chamkha A.J. [2018], "Unsteady MHD boundary layer flow of tangent hyperbolic two-phase nanofluid of moving stretched porous wedge", Int. Numer. J. Methods Heat Fluid Flow, 28(11):2567–2580.
13. Sadia S., Nahed B., Hossain M.A., Gorla R.S.R., Abdullah A.A.A. [2018], "Two-phase natural convection dusty nanofluid flow", Int. Heat J. Mass Transfer., 118:66–74.
14. Khan S.U., Shehzad S.A., Rauf A. & Ali A. [2018], "Mixed convection flow of couple stress nanofluid over oscillatory stretching sheet with heat absorption/generation effects", Results Phys., 8:1223–1231.
15. Prakash J., Sivab E.P., Tripathi D. & Kothandapani M. [2019], "Nanofluids flow driven by peristaltic pumping in occurrence of magnetohydrodynamics and thermal radiation", Mater. Sci. Semicond. Proc., 100:290–300.
16. Prakash J., Siva E.P., Tripathi D., Kuharat S. & Anwar O. [2019], "Peristaltic pumping of magnetic nanofluids with thermal radiation and temperature-dependent viscosity effects: modelling a solar magneto-biomimetic nanopump", Renew. Energy, 133: 1308–1326.
17. Akbar N.S., Huda A.B., Habib M.B. & Tripathi D. [2019], "Nanoparticles shape effects on peristaltic transport of nanofluids in presence of magnetohydrodynamics", Microsyst. Technol., 25:283–294.

11 An Interactive Injection Mold Design with CAE and Moldflow Analysis for Plastic Components

Sudhanshu Bhushan Panda and Antaryami Mishra

11.1 INTRODUCTION

Injection molding of plastic products is a complex process. In this technique, the quality product depends on the process characteristics, properties of polymer, mold construction, and so on. It is not an easy task to build a correct relation between the process and product quality. However, it is not productive to fix an issue in production by altering the process parameters and mold construction depending on a conventional experimentation technique. So, the logical tools for injection molding process have been in urgent need. Hence, the simulation programming of injection molding process evolved and has been effectively applied. With this simulation tool, an impersonation of the injection molding process measure is introduced, and it is much better to maintain a strategic distance from issues in the design stage than to fix problems in the production phase based on logical analysis of the process [1].

Advances in computer innovation have empowered enormous investigations utilizing programs for simulation based on a finite element method of analysis, for example, moldflow. Simulation programming is utilized broadly in the design of injection tools, due to effective prediction of the polymer melt flow to depict the mold-filling measure. Such analysis programs empower the mold filling, post-filling,

Email: sudhanshupanda@gmail.com; igit.antaryami@gmail.com

DOI: 10.1201/9781003216278-11

solidification, and cooling phases of the injection-molded part to be experimented on. Thus, it is conceivable to encapsulate the complexity of polymer melt behavior and progress in science in calculations which effectively model the filling process. In consequence, component manufacturability and process viability can be enhanced [2].

11.2 INJECTION MOLDING

In the cycle of injection molding, at first the mold is closed to ensure the geometry into which the molten plastic material is injected. The screw then activates and moves forward and a piston forces melted plastic into the mold cavity. This is referred as the filling or injection phase. The moment filling is achieved, pressure is maintained on the polymer melt and this is the beginning of the packing phase. The intention of the packing phase is to accumulate further plastic melt to compensate for part shrinkage as it cools and solidifies inside the cavity.

During the packing phase, the feeding point solidifies and the link to the mold cavity is effectively cut by the force applied by the polymer melt in the machine barrel. This denotes the start of the cooling process, in which the material keeps on dissipating heat until the part has enough mechanical rigidness to be ejected from the cavity. In the cooling phase, the screw begins to rotate and moves backward. The rotation facilitates plasticization of the polymer material and a fresh new charge of plastic melt is built up at the tip of the machine screw. At the time, the part is fairly solid, the mold begins to open and the component is ejected. Then the mold closes and the molding cycle starts again.

The process of injection molding is the most significant procedure for the creation of molded parts made of thermoplastics, thermosets, or elastomers in the automotive business as well as in all enterprises utilizing polymers (Figure 11.1). Despite the fact that injection molding tools are costly, the technology is favored because of the completely automated process along with the high part counting and the fantastic reproducibility of molding parts [3]. At present, about 30% of all thermoplastic material is produced by injection molding and the majority of all plastic handling articles are injection-molded. Perhaps the most progressive innovation to influence injection molding in recent years would be computerized implementations in modern manufacturing [4].

11.3 INJECTION MOLD DESIGN

The nature of an injection plastic part is influenced by numerous elements that incorporate geometric boundaries connected with the tool design, cooling configuration, and procedure boundary conditions, for example, molding conditions during the filling stage. To enhance the tool design and process conditions, increasingly the emphasis has been on computer-supported attributes, production, and engineering [5]. The computerized world is quickly progressing with the global concerns of automotive manufacturing. Mechanical development will empower greater levels of safety and accommodation [6]. Innovative work and highest-quality production are significant for vehicle manufacturers and their allied equipment providers. This calls for equipment testing that can offer sensible, viable reproduction for the whole scope of

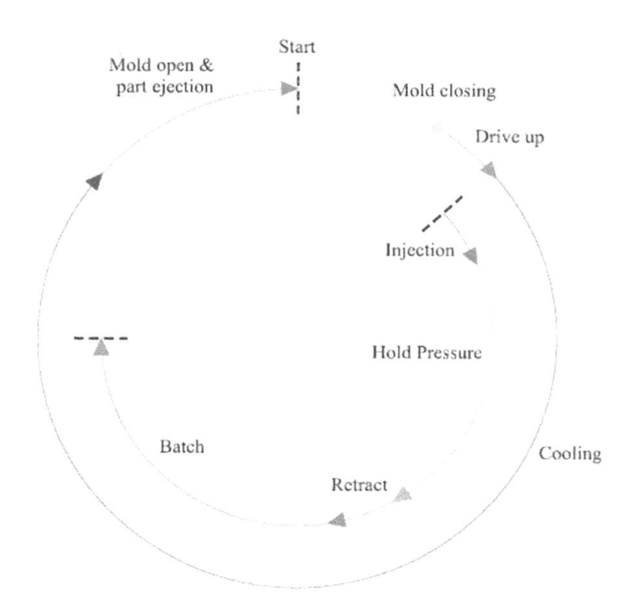

FIGURE 11.1 Injection molding process.

design integration, while permitting a quick account of verification and analysis of facts and figures in the early stage of product development [7, 8].

Techniques for computer-assisted engineering (CAE), from which computer-based production planning and mold design developed, are referred to as the most significant engineering tool in manufacturing, especially in the manufacturing sector. This rising role is firmly emphasized by the rapid advancement of the finite element method of analysis and modeling [9, 10].

Design for process planning and mold development for injection part manufacturing can profit by the joined-up utilization of information-based frameworks and process modeling. As of late, numerous organizations have been implementing computer-aided design and manufacturing (CAD-CAM) methods and information-related expert systems to improve and automate mold design and production activities. The overall plan of this information-based expert framework [9, 10] can be found in many industries at present. In connection with this framework, production planning as well as tool and die configuration capacities have been incorporated into an information-based master integration of computer-based manufacturing technology. It has a modular structure with all-around characterized incorporation of every module, giving smoothed-out information and data streams between different design stages. It comprises a mathematical and geometric module for creating, sending out, and bringing in the product geometrical calculations, a clear section for deciding the ideal shape, dimensions, and settling of spaces, a mechanical structure for planning the procedure grouping dependent on experimental guidelines and innovative boundaries, and an instrument structure module for structuring the mold and choosing a device of standard shape, and for getting ready projects for manufacturing of molds.

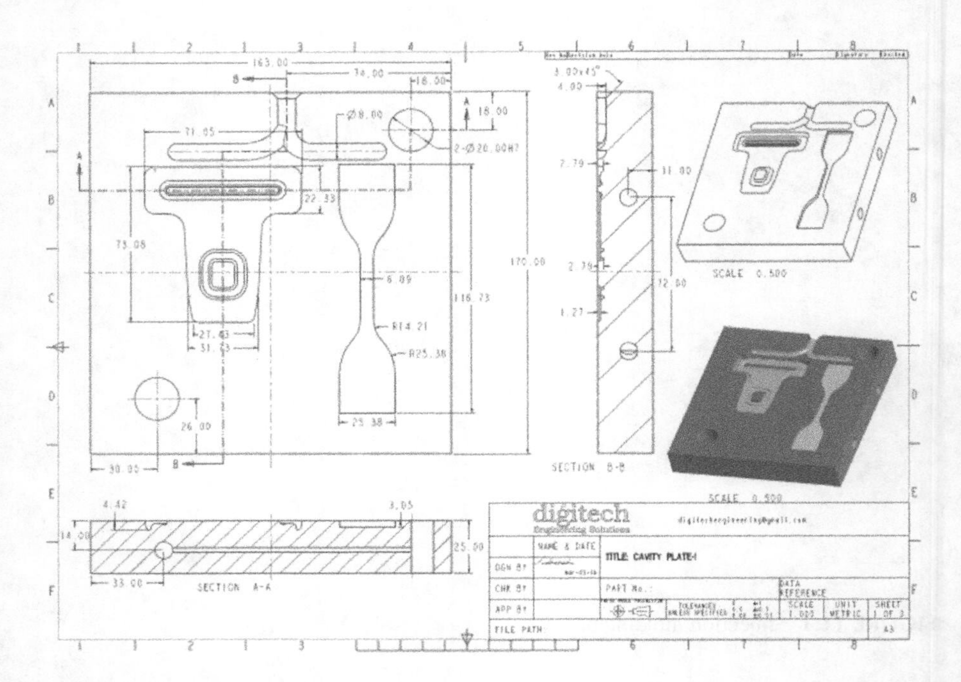

FIGURE 11.2 Injection mold design.

In this study the mold design generated integrating CAD, CAE, and moldflow is shown in Figure 11.2 [10–13].

Enhancement of molds and items using experimentation strategies, i.e., by tool and die modification and alterating the procedure boundaries in a progression of experiment moldings, is costly and tedious. In implementing logical oriented CAE tools for simulation and analysis, specialists can assess selective design plans and materials without genuinely investing in material and machine time [14, 15]. Computer-supported engineering has been created and applied to design product structure, shape, and preparatory conditions and to anticipating product defects in all designing and engineering fields. As the numerical examination programs have quickly improved in the most recent decade, numerical investigation is broadly utilized as a standard device to decide product design and geometrical shape, size calculations, and procedure conditions before genuine manufacturing [16, 17].

11.4 MOLDFLOW ANALYSIS

In this regard, the use of different strategies for CAE, including computer-assisted mold design and computer-regulated manufacturing, as well as coordinated part development, procedures and quick start of tool and die fabrication with the use of numerical analysis and simulation, for the most part selected component investigation and simulation should be referenced. At present, accessible computer method and different devoted program codes are especially created for delivering products and as a rule in close participation with the manufacturing units. In light of these turns of events, injection mold production, that traditionally had been the specialty of mold-making experts, now appears to be a genuine science-oriented design control [18]. During the last 10–15 years, CAE strategies, including computer-assisted process planning and mold design, have progressed to become among the prominent engineering instruments in manufacturing, especially in the manufacture of plastic parts. In manufacturing, process planning and injection mold design configuration can profit by a consolidated utilization of information-based frameworks and demonstrations of procedure. As of late, numerous organizations are implementing CAD-CAM procedures and information to master frameworks to improve fabrication capacity [10]. After materials and items have been chosen, a significant step forward is to advance the shape structure in order to improve the manufacturing procedure and limit conceivable molding defects [4]. During the last decade, numerous investigators have created CAD-CAE to shape tool and die design frameworks for plastic injection molding [19].

In this experiment, polyoxymethylene (POM) was taken as the material of investigation. The raw material data was input to the moldflow simulation platform, as noted in Table 11.1. The material has the trade named Delrin and the raw material manufacturer is DuPont.

In this experiment of moldflow analysis the melt and temperature were taken to be 210°C and 80°C, respectively. The maximum injection pressure was set at 140 MPa. The raw material density was taken as 1.42 g/cc and the shrinkage value 2.1%, as specified by the raw material manufacturer's data sheet. Manufacturers combine expertise in materials and processing with an understanding of design. Product planning, injection mold design, and optimum process characteristics were all assisted by process simulation, and were important for the success of plastic injection molding of the part [20, 21].

Trial-and-error methods take a significant amount of time in the product development process. But modeling and analysis at an early stage are increasingly being referred so as to reduce costs and save valuable time before manufacturing begins. It is important to differentiate precisely between cause and effect [22]. Computer

TABLE 11.1
Raw Material Data

Material	Polyoxymethylene
Grade	Delrin 100 AL NC010
Manufacturer	DuPont
Melt flow index (MFI)	MFI (1902.16) = 2.5 g/10 min
Fiber weight percent (%)	No fiber
Melt temperature (°C)	210
Mold temperature (°C)	80
Injection pressure (MPa)	140
Polymer density (g/cc)	1.42
Shrinkage (%)	2.1

simulation of the filling, packing, and cooling stages of polymer injection molding requires the values of the physical characteristics of the plastic and the raw material from which the mold is fabricated [17]. Moldflow simulation suggests judicious solutions for most of the complex parameters that lead to complications in the process of injection molding [23]. This procedure entails many design characteristics that need to be considered in an integrated approach. Tool design is a knowledge-intensive process. Numerous researchers worldwide have focused for a long time on tool design for polymer injection molding assisted by computers. Over the past 10 years, many studies have evolved as CAD-CAE tool design processes for polymer injection parts. Researchers have implemented CAD and CAE to develop accurate control for selection of injection molding process parameters [24]. The mathematical simulation is used as the first guideline model giving exact portrayals and patterns of the injection molding process cycle. In addition to the mathematical representation, a number of logical simulation patterns were also developed to validate the process characteristics of injection-molded parts [25].

The final part is generally imported to the software platform and specified with the plastic raw material data from the raw material manufacturer's database. The best gate or feeding point location is selected by the user to predict the result by simulation and analysis with the moldflow program. The material library is fed with rheological, mechanical, and thermal characteristics of materials. Engineers select values of injection molding process and choose the feeding point for the gating and runner systems. Further investigations are done particularly for the meltflow, fill duration, injection force, pressure decrease, melt front heat, weld-line occurrence, air trap locations, and quality of cooling [24].

The time for filling and material packing mainly depends on the computer-aided simulation, particularly with the aid of computational principles of fluid dynamics and simulation. Analysis of numerical technique provides directions to find convenient solutions to such logical problems. Such strategies will give estimated but acceptable troubleshooting for the issues, to predict injection molding defects and possible remedies. Numerical simulation based on numerical principles has developed along with studies of fluid mechanics, heat transfer, phenomena of transport techniques, and, obviously, rheology of polymer and its processing [25].

11.5 RESULTS AND DISCUSSION

The injection molding process involves polymer melt passing through a distribution system consisting of sprue, runner, and gates to independent cavities. An appropriate feed system is vital, particularly for a multiple impression mold since it mandates the filling behavior, prevents overpacking, eliminates defective molded components, and effectively enhances productivity [21]. The simulation results shown in Figure 11.3, with graphical representation for sprue pressure development at sprue opening, as shown in Figure 11.3(a), is the first gateway to mold cavity from the injection machine nozzle. This analysis displays the graph of pressure at sprue versus filling duration. It is useful to examine this output if any unexpected pressure increases at sprue at the time of filling. If the outcome profile of the pressure at the sprue is constant at the

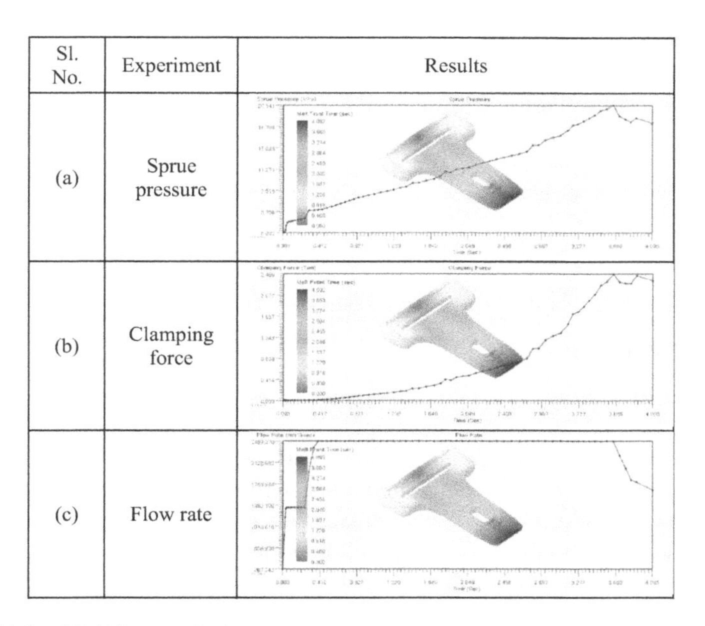

FIGURE 11.3 Moldflow analysis.

maximum specified pressure of injection, there is a chance of obstruction in melt flow or short shot molding.

The sprue is the part of the mold in which the molten plastic flows at first. Since it is in direct contact with the nozzle of an injection molding machine, it is susceptible to wear. Sprue brushing minimizes plastic material loss, sufficiently tapered to smoothly separate products from the mold and without undercuts on the inner surface [26].

This result, shown in Figure 11.3(b), shows a graphical representation of the clamping tonnage compared with filling duration. The resulting value is the essential clamping tonnage at the time of molding production in lieu of the force of that molding machine's output. This outcome is meant to recognize the molding flash issues. If the estimated clamping pressure is greater than 70% of the higher side of the clamping force of the machine, polymer melt may be compressed by the outer periphery of the impression that causes it to flash.

Figure 11.3(c) shows the flow rate result at the sprue opening compared with filling duration of the analysis. In general, the initial filling stage is regulated by the rate of flow pressure set by the operator. Hence, in this outcome, the flow rate generally remains at the parameter specified in the molding process condition. If the experimenting flow rate differs, then the maximum allowed injection pressure should be verified.

The moldflow simulation is carried out with various parameters as inputs. The filling time is maintained at 4 seconds. The output parameters for injection molding of the said polymer are noted in Table 11.2. These output values may refer in injection molding machine during production to get defect-free parts with less effort.

TABLE 11.2
Moldflow Simulation Results

Ejection temperature (°C)	115
Freeze temperature (°C)	130
Filling time (seconds)	4.000
Packing time (seconds)	13.700

11.6 CONCLUSION

In this competitive strategy of delivering products that are right the first time and with the lowest lead time for product development, integration of CAD with CAE tools such as moldflow simulation programs inevitably are placed in the first category of market competition. CAM also plays an important role in the manufacture of injection molds directly from CAD-developed mold designs. Advances in computer technology have enabled precise prediction of injection molding characteristics using simulation tools based on finite element method of analysis like moldflow. Injection design aided by moldflow simulation is crucial for the success of plastic injection molding products. Computer-supported design, analysis, and interactive manufacturing technology significantly improve the quality of parts and hassle-free production systems.

REFERENCES

1. H. Zhou, "Computer Modeling for Injection Molding: Simulation, Optimization, and Control," John Wiley, pp. 03–57, 2013.
2. S. I. Shaharuddin, M. Salit, E. S. Zainudin, "A review of the effect of molding parameters on the performance of polymeric composite injection molding," Turkish Journal of Engineering & Environmental Sciences, Vol. 30(1), pp. 23–34, 2006.
3. J. Rowe, Advanced Materials in Automotive Engineering, Woodhead Publishing, 2012.
4. M. Holmes, "Fibre-reinforced composites used for cost effective aerospace window frames," Reinforced Plastics, Vol. 58(1), p. 08, 2014.
5. A. Kumar, P. S. Ghoshdastidar, M. K. Muju, "Computer simulation of transport process during injection mold-filling and optimization of the molding conditions," Journal of Materials Processing Technology, Vol. 120, pp. 438–449, 2002.
6. J. Reiner, R. Pletziger, J. Buss, "The true value of autonomous driving," Automotive Manager Report, Vol. 18(03), pp. 07–11, 2015.
7. G. Wohlke, E. Schiller, "Digital planning validation in automotive industry," Computers in Industry, Vol. 56(4), pp. 393–405, 2005.
8. J. Balic, "Intelligent CAD/CAM systems for CNC programming: An overview," Advances in Production Engineering & Management, Vol. 1(1), pp. 13–22, 2006.
9. D. D. Sharmah, "Autotech Review," Vol. 2(7), pp. 01–63, 2013.
10. M. Tisza, "Recent development trends in sheet metal forming," International Journal of Microstructure and Materials Properties, Vol. 8(1/2), pp. 125–140, 2013.
11. G. Boothroyd, "Product design for manufacture and assembly," Computer-Aided Design, Vol. 26(7), pp. 505–520, 1994.

12. A. A. Fagade, D. O. Kazmer, "Early cost estimation for injection molded parts," Journal of Injection Molding technology, Vol. 4(3), pp. 97–106, 2000.
13. A. E. Tekkaya, "State-of-the-art of simulation of sheet metal forming," Journal of Materials Processing Technology, Vol. 103(1), pp. 14–22, 2000.
14. A. Marcilla, A. Odjo-Omoniyi, R. Ruiz-Femenia, J. C. Garcıa-Quesada, "Simulation of the gas-assisted injection molding process using a mid-plane model of a contained-channel part," Journal of Materials Processing Technology, Vol. 178, pp. 350–357, 2006.
15. J. Hassan, J. Balzulat, T. Carr, S. Tylko, "The electronic belt-fit test device (eBTD): Activity update," In Proceedings of the 20th International Technical Conference on the Enhanced Safety of Vehicles (ESV), Vol. 7, pp. 1–13, 2007.
16. K. Jin, T. Kim, N. Kim, B. Kim, "Process chain analysis of the dimensional integrity in a metal-insert polymer smart phone base plate — from die casting to polymer injection molding," Journal of Mechanical Science and Technology, Vol. 29(4), pp. 1703–1713, 2015.
17. S. Kulkarni, D. J. Edwards, E. A. Parn, C. Chapman, C. O. Aigbavboa, R. Cornish, "Evaluation of vehicle lightweighting to reduce greenhouse gas emissions with focus on magnesium substitution," Journal of Engineering, Design and Technology, Vol. 16(6), pp. 869–888, 2018.
18. E. Mangino, J. Carruthers, G. Pitarresi, "The future use of structural composite materials in the automotive industry," International Journal of Vehicle Design, Vol. 44(3/4), pp. 211–232, 2007.
19. I. Matin, M. Hadzistevic, J. Hodolic, D. Vukelic, D. Lukic, "A CAD/CAE-integrated injection mold design system for plastic products," The International Journal of Advanced Manufacturing Technology, Vol. 63(5–8), pp. 595–607, 2012.
20. A. Roberts, M. Partain, S. Batzer, D. Renfroe, "Failure analysis of seat belt buckle inertial release," Engineering Failure Analysis, 14(6), pp. 1135–1143, 2007.
21. S. Salunkhe, S. Teraiya, H. M. A. Hussein, S. Kumar, "Smart system for feature recognition of sheet metal parts: A review," Innovative Design, Analysis and Development Practices in Aerospace and Automotive Engineering (I-DAD 2018), Lecture Notes in Mechanical Engineering, Vol. 2, pp. 535–549, 2018.
22. P. Nambiar, G. Padma, S. Thakore, "Towards Responsible Use of Plastics Reduce, Reuse, Recycle," Centre for Environment Education, India, pp. 01–10. ISBN: 978-93-84233-58-7, 2018.
23. F. Henriksson, J. Detterfelt, "Production – as seen from product development: A theoretical review of how established product development process models address the production system," Proceedings of NordDesign 2018: Design in the Era of Digitalization, Linkoping, Sweden, pp. 01–11, 2018.
24. S. B. Panda, N. C. Nayak, A. Mishra, "Characterization and optimization of tool design of an injection molded part through mold-flow analysis", Lecture Notes on Data Engineering and Communications Technologies 37, Advances in Data Science and Management Proceedings of ICDSM 2019, Vol. 37, pp. 453–461, 2019.
25. R. Zhang, Z. Shao, J. Lin, "A review on modelling techniques for formability prediction of sheet metal forming," International Journal of Lightweight Materials and Manufacture, Vol. 1, pp. 115–125, 2018.
26. I. Staffell, D. Scamman, A. Velazquez Abad, P. Balcombe, P. E. Dodds, P. Ekins, N. Shah, K. R. Ward, "The role of hydrogen and fuel cells in the global energy system," Energy & Environmental Science, Vol. 12, pp. 463–491, 2019.

12 Study of Collaborative Filtering-Based Personalized Recommendations

Quality, Relevance, and Timing Effect on Users' Decision to Purchase

Darshana Desai

Email: darshana.j.desai@gmail.com

DOI: 10.1201/9781003216278-12

12.1 INTRODUCTION

Personalization is used as an effective tool in virtual markets on e-commerce websites with recommendations of information, product, or services to reduce users' cognitive load. Users' demographic information, purchase habits, likes, social interactions, geographic locations, and behavior on the internet are continuously gathered to serve them better by e-commerce sites. Real-time personalized recommendations are presented to cater to the diverse needs of individuals by understanding implicit likes, browsing of websites, behavioral intentions like time spent on the page, number of visits to pages, frequency of item viewed, search and purchase history with real-time analytics. To generate personalized recommendations and produce highly relevant content, detailed user information is created on a real-time basis upon constant monitoring of users' activities which have raised privacy concerns and trust issues in e-commerce websites. The personalization process needs more data, monitoring users' activities in the right context to generate highly relevant personalized recommendations, whereas users experience more privacy concerns about the tracking of their activities and interactions. Do-not-track initiatives and limiting data collection by providing users with control regarding data collection reduce privacy concerns in users [1, 2].

Recently users' privacy concerns have been notably raised and play major role in developing trust before purchasing on e-commerce websites. Highly personalized recommendation and user behavior tracking may raise concerns about privacy and security. Personalization strategies have been extensively used over a decade to lower the cognitive load on the user and influence towards high returns in business but there has been little focus on customer data privacy and their concerns about information disclosure. With a view to addressing the gap, our research has explored personalized recommendations used by the e-commerce website Amazon.com on user purchase behavior with respect to privacy concerns and trust in the website. The chapter is organized as follows: Section 12.2 describes recommendations on Amazon.in and other contexts of personalized recommendation attributes of relevance, quality and timing, trust, privacy concerns, satisfaction, and their effect on purchase behavior. Section 12.3 presents the proposed research framework and hypotheses, describing the research methodology adopted to address the research question and analysis. Subsequently, Section 12.4 presents analysis of the results. Finally, Section 12.5 discusses the results and findings with hypothesis-testing results and proposes the future scope of research.

12.2 RELATED WORK

Personalized recommendations produced on e-commerce websites on a real-time basis is a continuous process addressing users' requirements with highly relevant information [3]. Users' needs are identified implicitly by analyzing their search history, website clicks, purchase history, demographics, and likes and explicitly producing customization options and choices of filters [4–7]. Research suggests that recommendation services reduce users' cognitive load, producing ease of use, perceived usefulness, and enjoyment of the shopping experience, subsequently leading to positive purchaser behavior and revisit to the e-commerce websites [7–9]. Users experience higher satisfaction and more trust with greater interaction with an e-commerce website, show positive behavioral intentions, and purchase [3, 10]. In contrast, users experience higher privacy concerns and lower satisfaction with recommendations associated with higher financial risks [11, 12]. Research also suggests that recommendations of sensitive products generate higher privacy concerns in users who are likely to experience lower trust in the e-commerce websites, having a negative effect on behavioral intentions such as revisiting and purchasing.

12.2.1 Personalized Recommendations on Amazon.in Website

Personalized recommendation systems adopt different approaches and techniques to learn about users' preferences by analyzing their behavioral data like collaborative filtering, user profiling, content-based and hybrid model-based recommendations [4, 13]. Research has found significant performance improvement in business and customer experience with personalized recommendations in e-commerce websites like Amazon.com which use a collaborative filtering technique to generate product recommendations by analyzing the purchase behavior of like-minded people and suggesting related items which are frequently bought together [14, 15].

Frequent interaction with e-commerce websites and increased use of recommendation services produce increased satisfaction, higher trust, and greater purchase intentions. However trust in the recommendation is strongly associated with users' desire to avail themselves of personalization services [10] and greater relevance of personalized recommendations to users' implicit need. The accuracy of recommendations induces more satisfaction in users, instigating positive behavioral intentions like revisiting the website and purchasing [3, 7, 11]. Recommendations made to provide personalized product listings have the potential to influence users' decision making. The Amazon recommendation system uses a collaborative filtering algorithm to produce recommendations, which generates continuous feedback with increased user interaction using behavioral tracking. An item-based collaborative filtering method using popular and frequently used items creates a bias which affects users' preferences on a real-time basis.

The increase in e-commerce website usage and addressing diverse needs of users have been the focus of research on personalized recommendation systems. In order to design effective personalized recommendation algorithm systems, information is gathered about users by explicit and implicit means. Users' explicit information is gathered from customization options given to users for efficient search of product, demographic profiles, purchase history, and implicit information like purchase behavior and click stream data. Specifically an item-to-item collaborative filtering technique is used to predict customers' preferences on the basis of like-minded customer interactions and preferences. A recommendation algorithm reviews visitors' recent purchase history and draws up a list of related items. Recommended items with the highest ranking across the list are listed and repeated. A collaborative filtering technique is used to personalize users' experience through recommendations generating product information tailored to users' interests, leveraging the experiences of other users with similar profiles. Traditionally, the technique is used in e-commerce platforms to drive sales by converting targeted suggestions to purchases. The technique has rendered more favorable results than blanket advertisements, and is more purposeful toward customizing the user's experience [16]. The quality of recommendation depends on the recent history of purchase, timing of recommendation, and relevance to users' current needs and interests. Users' needs and interests change frequently, and this is difficult to address with real-time analytics. The current research focuses on users' purchase behavior with the recommender system on the online e-commerce shopping website Amazon.in.

12.3 RESEARCH FRAMEWORK AND HYPOTHESES

Research in personalization has shown a significant relationship between information personalization and users' intentions to revisit the website. Our research framework proposes the interrelation of personalized recommendations, privacy concerns, trust, and behavioral intentions of user-liked purchases in e-commerce websites.

12.3.1 Personalized Recommendations, Privacy Concerns

To provide personalized content, applications require users' personal information [17, 18] and understand their implicit need by observing their interaction with the website [19], their purchase behavior, and what they like; this has the potential to be an invasion of privacy. Individuals might have higher privacy concerns about online personalization if they are not aware of the intentions behind content personalization which reduces trust in social networking sites. Users experience higher privacy concerns when the personalization process uses their information without their consent and if they have negative feelings about personalization.

H1: Users exhibit higher privacy concerns with personalized recommendation.

12.3.2 Personalized Recommendations and Trust

Recommendations produced for users on a real-time basis identifying their implicit needs are done with a collaborative filtering technique used on the Amazon.in website. Recommendations with high relevance to users' implicit need induce trust amongst users and they are likely to revisit the website. Users experience higher satisfaction with highly relevant personalized information, perceived ease of use, and usefulness of the recommendation. Explicit personalization or customization choices offered to users produces a sense of control and involvement in the personalization process which leads to higher satisfaction and joyful interaction with personalized e-commerce sites [3], which in turn leads to higher trust in the ecommerce recommendation. So our work hypothesizes:

H2: Users experience more trust with highly relevant, good-quality and timely personalized recommendations.

12.3.3 Privacy Concerns and Trust

Personalization of websites reduces cognitive efforts by users to search for information and they feel higher satisfaction with more relevant personalized information [3], eventually developing trust in websites. Users with greater privacy concerns need more control over personalization to develop trust in social networking websites. Users' privacy concerns lead to trust building upon fulfillment. Personalized recommendation generates cognitive benefits to users at the cost of information disclosure. In the process of achieving personalization, websites need to collect more information about users' personal and behavioral characteristics like interests and implicit needs. This increases privacy concerns when privacy risks are associated, particularly if users are wary of personalized content [20]. So research postulates:

H3: Users exhibit higher trust with addressed privacy concerns.

12.3.4 Privacy Concerns and Purchase Intentions

Privacy concern is the user's builtin desire to control procurement and information usage upon sharing or acquired through website interaction and transactions. Users' privacy concerns are related to preservation of their anonymity and are strongly associated with their control of personalization and their explicit customization through settings and preferences. Earlier research has found that information control is a major factor in users' privacy concerns in an online environment [21]. Users have high privacy concerns when they perceive there to be a high risk and threat to their privacy upon information sharing and less control over personalization settings. Similarly, users will have less concern about privacy when provided with the option to control the visibility of posted content. In contrast, users experience lower privacy risks when provided with higher control of exposure to privacy policies and sharing personal information [22]. So the research postulates:

H4: Users exhibit higher purchase intention when their privacy concerns are addressed on e-commerce sites.

12.3.5 Trust and Purchase Intentions

Trust is users' intrinsic feeling and willingness to be vulnerable to the actions with expectations and when their expectations are addressed [23]. Users who have control over information flow, profile visibility, and protection of their personal information are more likely to develop higher satisfaction and trust in e-commerce websites. Highly relevant personalized content offerings on websites, like a targeted advertisement, recommendation, location tracking, and suggestions based on geography, increase users' satisfaction and induces an intrinsic feeling of trust which motivates users to purchase. So, we propose:

H5: Users' purchase intention is highly associated with their trust.

12.4 RESEARCH METHOD

12.4.1 Data Collection and Sampling

Research focuses on the purchase intentions of Amazon.in e-commerce website users who have been using this site for more than 2 years. The research used the online survey-based method for data collection. All the constructs identified from previous studies – personalized recommendations, trust, privacy concerns, satisfaction, and purchase intentions – were measured on a 5-point Likert scale.

12.4.2 Measurement Model

The population of data collection was respondents from India who purchased online on Amazon.in e-commerce websites and experienced personalization in the form of the personalized recommendations of products, services, or information directly or indirectly. A total of 465 valid responses were collected from 500 responses through data preprocessing. Data cleaning was done after the removal of noisy, incomplete, and inconsistent data. Responses with a standard deviation below 0.30 were removed to enable final valid responses for further data analysis. Factors were identified and confirmed using exploratory and confirmatory factor analysis methods using SPSS 20.0. The structural equation modeling (SEM) technique was used to identify the model fit of the proposed model.

KMO and Bartlett's test

Kaiser–Meyer–Okin measure of sampling adequacy		0.848
Bartlett's test of sphericity	Approx. chi-square	2819.144
	Df	91
	Sig.	0.000

Kaiser–Meyer–Okin (KMO) and Bartlett's test results have a value of 0.848 which is above 0.7 and shows sample size adequacy of data collection. Exploratory factor analysis techniques discovered five factors: personalized recommendations, trust, privacy concerns, satisfaction, and purchase intentions through extraction method maximum likelihood. Table 12.1 shows four factors with 56.78% of the cumulative load.

Cronbach's alpha coefficient values of construct items were observed in the range of 0.70–0.90, showing higher internal consistency of constructs and consistency of scale of questionnaire items. Construct item factor loading values were identified above 0.6 in Table 12.2, which shows satisfactory factor loading with higher construct validity and reliability.

Table 12.3 is the factor correlation matrix showing the interrelation of the factors identified after exploratory factor analysis.

12.4.2.1 Structural Equation Modeling

The research model was tested with a structural equation model using AMOS 21.0. The model fit indices v2/df = 1.772, goodness of fit index (GFI) = 0.961, adjusted goodness of fit index (AGFI) = 0.94, normed fit index (NFI) = 0.95, comparative fit index (CFI) = 0.98, RMSEA = 0.039 were achieved, indicating that it is a good model. All the indices for model fit in structural equation modeling indicate that it is proposed that a model should have a good fit.

TABLE 12.1
Total Variance

Factor	Initial eigenvalues			Extraction sums of squared loadings			Rotation sums of squared loadings[a]
	Total	% of variance	Cumulative %	Total	% of Variance	Cumulative %	Total
1	4.843	34.595	34.595	4.024	28.741	28.741	3.041
2	2.428	17.341	51.936	1.935	13.822	42.563	3.328
3	1.156	8.255	60.191	1.253	8.948	51.512	2.999
4	1.074	7.670	67.860	0.739	5.277	856.788	2.808
5	638	4.554	72.414				
6	0.585	4.176	76.589				
7	0.561	4.008	80.597				
8	0.539	3.850	84.447				
9	0.452	3.225	87.672				
10	0.401	2.886	90.539				
11	0.385	2.748	93.286				
12	0.369	2.635	95.921				
13	0.325	2.321	98.242				
14	0.246	1.758	100.000				

Extraction method: Maximum likelihood.

[a] When factors are correlated, sums of sqaure loadings cannot be added to obtain a total variance.

Table 12.4 displays the standardized path coefficients, variance explained (R^2), and path significances values for the path as per the proposed hypothesis, and all hypotheses are supported except for correlation of trust and satisfaction. As with variance explained (R^2), R^2 values of purchase intentions and privacy concerns are 0.17 and 0.45 which is above 0.3 and shows a good research model.

12.5 RESULTS AND DISCUSSION

The SEM result in Figure 12.1 shows that users experience higher privacy concerns with an increase in personalized recommendations and feel satisfied with the high relevance of recommendations to their implicit need, leading to higher satisfaction. All the hypotheses proposed in the model satisfy users' trust dependency on privacy concerns in an e-commerce website. Users are more likely to purchase from e-commerce websites when their privacy concerns are addressed properly by providing them with more control. In contrast, users experience less trust with higher privacy concerns. The result also shows that the willingness of the user to purchase is not affected by trust but is highly affected by satisfaction level with a personalized recommendation.

TABLE 12.2
Pattern Matrix[a]

	Factor			
	1	2	3	4
ECIP2	0.793			
ECIP3	0.764			
ECIP4	0.724			
ECIP5	0.653			
ECIP1	0.608			
ECPC2		0.862		
ECPC1		0.702		
ECPC3		0.692		
ECPC4		0.612		
ECPUR2			0.865	
ECPUR1			0.753	
ECPUR3			0.519	
ECTrust2				0.975
ECTrust1				0.701

Extraction method: Maximum likelihood.
Rotation method: Promax with Kaiser normalization.
[a] Rotation converged in six iterations.

TABLE 12.3
Factor Correlation Matrix

Factor	1	2	3	4
1	1.000	0.228	0.308	0.369
2	0.228	1.000	0.569	0.510
3	0.308	0.596	1.000	0.448
4	0.369	0.510	0.448	1.000

Extraction method: Maximum likelihood.
Rotation method: Promax with Kaiser normalization.

12.6 CONCLUSIONS AND FUTURE SCOPE OF RESEARCH

This research is a qualitative study on users' purchase intentions in Amazon.in on an e-commerce website. Users' privacy concerns and trust with personalized recommendations play a significant role in developing trust and positive behavioral intentions to purchase from the e-commerce websites. Highly relevant recommendations addressing users' implicit needs generate higher satisfaction and develop trust in the website. The quality of personalized recommendations, their

TABLE 12.4
Hypothesis Testing Results

	Estimate	S. E.	C.R.	P-label
Trust ← Personalized recommendation	0.427	0.042	10.103	***
Privacy concerns ← Personalized recommendation	0.039	0.040	0.966	0.334
Privacy concerns ← Trust	0.535	0.039	13.834	***
Purchase intention ← Privacy concerns	0.546	0.040	13.699	***
Purchase intention ← Trust	.174	0.038	4.538	***

*** Parametric constraints.

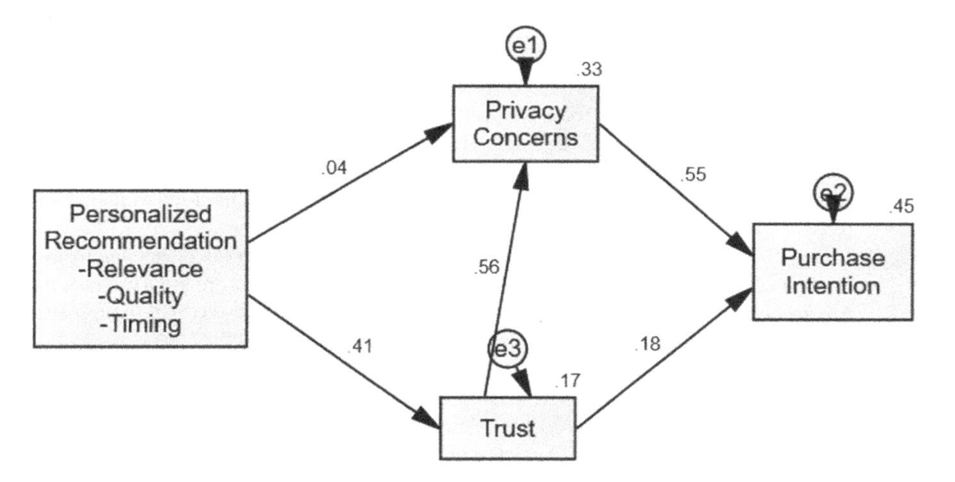

FIGURE 12.1 Structural equation modeling results.

relevance and timeliness play a significant role in developing trust in users and further motivate them to purchase from the e-commerce website. Users' purchase intentions are affected when their privacy concerns are addressed and satisfaction is increased with perceived ease of use and the usefulness of highly relevant personalized recommendation. However the results show that users' privacy concerns are not dependent on personalized recommendations; users may experience privacy concerns if their information is used without their consent or prior information for commercial use or any business benefit.

The survey-based research methodology adopted for this study can be further studied with regard to users' purchase intentions in a controlled lab environment with live interactions with personalized recommendations on other e-commerce websites, and in a comparative study of users' responses.

REFERENCES

1. Toch, E., Wang, Y. & Cranor, L.F. (2012). Personalization and privacy: a survey of privacy risks and remedies in personalization-based systems. *User Modeling and User-Adapted Interaction*, 22, 203–220. https://doi.org/10.1007/s11257-011-9110-z
2. Wang, Y. & Kobsa, A. (2007). Respecting users' individual privacy constraints in web personalization. In: Conati, C., McCoy, K. & Paliouras, G. (eds) User Modeling 2007. Lecture Notes in Computer Science, vol. 4511. Springer, Berlin. https://doi.org/10.1007/978-3-540-73078-1_19
3. Desai, D. (2019). An empirical study of website personalization effect on users intention to revisit e-commerce website through cognitive and hedonic experience. In: Balas, V., Sharma, N. & Chakrabarti, A. (eds) Data Management, Analytics, and Innovation. Advances in Intelligent Systems and Computing, vol. 839. Springer. https://doi.org/10.1007/978-981-13- 1274-8_1
4. Desai, D. & Kumar, S. (2015). Web personalization: A perspective of design and implementation strategies in websites. Khoj: Journal of Management Research and Practices. ISSN No: 0976-8262.
5. Taylor, D.G., Davis, D. & Jillapalli, R. (2009). Privacy concern and online personalization: The moderating effects of information control and compensation. Electronic Commerce Research, 9, 203–223. DOI:10.1007/s10660-009-9036-2
6. Chellappa, R.K. & Sin, R.G. (2005). Personalization versus privacy: An empirical examination of the online consumer's dilemma. Information Technology Management, 6, 181–202. https://doi.org/10.1007/s10799-005-5879-y
7. Liang, T.-P., Chen, H.-Y., Du, T., Turban, E. & Li, Y. (2012). Effect of personalization on the perceived usefulness of online customer services: A dual-core theory. *Journal of Electronic Commerce Research*, 13(4), 275–288. www.ecrc.nsysu.edu.tw/liang/paper/2/79 Effect of Personalization on the Perceived (JECR, 2012).pdf
8. Desai, D. (2016). A study of personalization effect on users' satisfaction with ecommerce websites. Sankalpa: Journal of Management and Research. ISSN No. 2231-1904.
9. Mahrous, A.A. (2011). Antecedents of privacy concerns and their online actual purchase consequences: a cross-country comparison. International Journal of Electronic Marketing and Retailing, 4(4), 248–269.
10. Dabholkar, P.A. & Sheng, X. (2012). Consumer participation in using online recommendation agents: effects on satisfaction, trust, and purchase intentions. The Service Industries Journal, 32(9), 1433–1449. https://doi.org/10.1080/02642069.2011.624596
11. Stevenson, D. & Pasek, J. (2015). Privacy concern, trust, and desire for content personalization. In: The 43rd Research Conference on Communication, Information and Internet Policy. http://dx.doi.org/10.2139/ssrn.2587541
12. Barth, S. & de Jong, M.D.T. (2017). The privacy paradox – investigating discrepancies between expressed privacy concerns and actual online behavior – a systematic literature review. Telematics and Informatics, 34(7), 1038–1058. https://doi.org/10.1016/j.tele.2017.04.013
13. Zhang, S., Yao, L., Sun, A. & Tay, Y. (2019). Deep learning-based recommender system: a survey and new perspectives. ACM Computing Surveys (CSUR), 52(1), 1–38.
14. Konstan, J.A., Riedl, J., Konstan, J.A. & Riedl, J. (2012). Recommender systems: from algorithms to user experience. User Modeling and User Adapted Interchanges, 22, 101–123. https://doi.org/10.1007/s11257-011-9112-x
15. Pu, P., Chen, L. & Hu, R. (2011). A user-centric evaluation framework for recommender systems. In: Proceedings of the Fifth ACM Conference on Recommender Systems (ACM), 157–164.

16. Linden, G., Smith, B. & York, J. (2003). Amazon.com recommendations: item-to-item collaborative filtering. IEEE Internet Computing, 7(1), 76–80.
17. Gupta, A. & Dhami, A. (2015). Measuring the impact of security, trust, and privacy in information sharing: A study on social networking sites. Journal of Direct, Data and Digital Marketing Practice, 17(1), 43–53. https://doi.org/10.1057/dddmp.2015.32
18. Shin, D. (2010). The effects of trust, security, and privacy in social networking: a security-based approach to understand the pattern of adoption. Interacting with Computers, 22(5), 428–438. https://doi.org/10.1016/j.intcom.2010.05.001
19. Desai, D. (2019). Personalization aspects affecting users' intention to revisit social networking site. International Journal of Trends in Scientific Research and Development (IJTSRD), 4(1), 612–621. www.ijtsrd.com/papers/ijtsrd29631.pdf
20. Stevenson, D. & Pasek, J. (2015). Privacy concern, trust, and desire for content personalization. In: The 43rd Research Conference on Communication, Information and Internet Policy. http://dx.doi.org/10.2139/ssrn.2587541
21. Sheehan, K.B. & Hoy, M.G. (1999). Flaming, complaining, abstaining: how online users respond to privacy concerns. Journal of Advertising, 28(3), 37–51.
22. Mohamed, N. & Ahmad, I.H. (2012). Information privacy concerns, antecedents and privacy measure use in social networking sites: evidence from Malaysia. Computers in Human Behavior, 28(6), 2366–2375. https://doi.org/10.1016/j.chb.2012.07.008
23. Dwyer, C., Hiltz, S.R. & Passerini, K. (2007). Trust and privacy concerns within social networking sites: a comparison of Facebook and MySpace. In: Americas Conference on Information Systems, Proceedings of the Thirteenth Americas Conference on Information Systems, Keystone, 9–12 August, Colorado, USA, p. 339. https://aisel.aisnet.org/amcis2007/339

Index

Q

R

S

T

V